AF389149

Wind Turbine Aerodynamics II

Wind Turbine Aerodynamics II

Editor

Wen Zhong Shen

MDPI • Basel • Beijing • Wuhan • Barcelona • Belgrade • Manchester • Tokyo • Cluj • Tianjin

Editor
Wen Zhong Shen
Yangzhou University
China

Editorial Office
MDPI
St. Alban-Anlage 66
4052 Basel, Switzerland

This is a reprint of articles from the Special Issue published online in the open access journal *Applied Sciences* (ISSN 2076-3417) (available at: https://www.mdpi.com/journal/applsci/special_issues/wind_turbine_aerodynamics_II).

For citation purposes, cite each article independently as indicated on the article page online and as indicated below:

LastName, A.A.; LastName, B.B.; LastName, C.C. Article Title. *Journal Name* **Year**, *Volume Number*, Page Range.

ISBN 978-3-0365-2472-6 (Hbk)
ISBN 978-3-0365-2473-3 (PDF)

Cover image courtesy of Prof. Dr. Wen Zhong Shen

Contents

About the Editor

Wen Zhong Shen (Dr) serves as Professor in Wind Energy at the College of Electrical, Energy and Power Engineering, Yangzhou University, China, since 2021. He obtained his B.Sc. degree in Mathematics at Wuhan University, China, in 1988, and M.Sc. and Ph.D. degrees in fluid mechanics at Paris-Sud University, France, in 1989 and 1993, respectively. He worked as a post-doc in LIMSI/CNRS (Centre National de la Recherche Scientifique) in 1993–1996 and subsequently held appointments as Assistant Research Professor, Associate Professor and Full Professor at the Department of Wind Energy, Technical University of Denmark (DTU), in 1996–2021. His research areas cover the fields of wind turbine aerodynamics, aero-acoustics, computational fluid dynamics, wind turbine airfoil/rotor design, and wind farm optimization.

Preface to "Wind Turbine Aerodynamics II"

As the pioneer of renewable energy, wind energy is developing very quickly all over the world. To reduce the levelized cost of energy (LCOE), the size of a single wind turbine has been significantly increased and will continue to increase further in the near future. This tendency requires the further development and validation of design and simulation models. This Special Issue "Wind Turbine Aerodynamics II" is a collection comprising numerous important works addressing the aerodynamic challenges appearing in such a development.

Wen Zhong Shen
Editor

Editorial

Special Issue on Wind Turbine Aerodynamics II

Wen Zhong Shen [1,2]

1   College of Electrical, Energy and Power Engineering, Yangzhou University, Yangzhou 225127, China;
     wen_zhong_shen@outlook.com
2   Department of Wind Energy, Technical University of Denmark, Nils Koppels Alle, Building 403,
     2800 Lyngby, Denmark

Citation: Shen, W.Z. Special Issue on Wind Turbine Aerodynamics II. *Appl. Sci.* **2021**, *11*, 8728. https://doi.org/10.3390/app11188728

Received: 15 September 2021
Accepted: 16 September 2021
Published: 18 September 2021

Publisher's Note: MDPI stays neutral with regard to jurisdictional claims in published maps and institutional affiliations.

1. Introduction

To alleviate global warming and reduce air pollution, the world needs to rapidly shift towards renewable energy. As the pioneer of renewable energy, wind energy is developing very fast all over the world. In order to capture more energy from the wind and reduce the levelized cost of energy (LCOE), the size of a single wind turbine has recently increased to 16 Mega-Watt (MW) [1], and will be increased further in the near future. Big wind turbines and their associated wind farms have advantages, but also challenges in all wind energy sciences, including wind turbine aerodynamics. The typical effects are mainly related to the increases in Reynolds number, in blade flexibility, and possibly in wind turbine noise. This Special Issue collects a number of important works addressing these aerodynamic challenges. Aerodynamics of wind turbines is a classic concept, and is the key for wind energy development, as all other wind energy sciences rely on the accuracy of its aerodynamic models. There are also several Special Issues on wind turbine aerodynamics. This guest editor edited a Special Issue in Renewable Energy on aerodynamics of offshore wind energy systems and wakes in 2014 [2], which collected state-of-the-art research articles on the development of offshore wind energy, and a Special Issue in Applied Sciences on aerodynamics in 2019 [3], which collected various important aerodynamics problems.

2. Current Status in Wind Turbine Aerodynamics

In the context introduced above, this Special Issue was to collect latest research articles on various topics related to wind turbine aerodynamics, which includes Wind turbine design concepts, Tip loss correction study, Wind turbine acoustics modelling, and Vertical axis wind turbine concept. A summary of the collected papers is given below in the order mentioned above.

There are also three papers dealing with Wind turbine design concepts. Sun et al. [4] presented a coned rotor concept with different conning configurations, including special cones with three segments. The authors made the analysis based on the DTU-10 MW reference rotor [5] and found that the different force distributions of upwind and downwind coned configurations agree well with the distributions of angle of attack, which are affected by the blade tip position and the cone angle, and the most upwind and downwind cones have a thrust difference up to 8% and a torque difference of up to 5%. The coned rotor concept has potential to be used for super-large wind turbines. The influence of tilt angle on aerodynamic performance of the virtual NREL 5 MW wind turbine [6] was studied by Wang et al. [7]. It was found that the change in tilt angle results in changing the angle of attack on wind turbine blade, which affects the thrust and power of the wind turbine, and the aerodynamic performance of the wind turbine is best when the tilt angle is about 4°. Subsequently, the effects of wind shear were also studied for the turbine with a tilt angle of 4°, and it was found that wind shear will cause the thrust and power of the wind turbine to decrease. Yang et al. [8] experimentally studied the effect of Gurney flaps on the performance of a wind turbine airfoil (DTU-LN221 airfoil [9]) under different turbulence levels (T.I. of 0.2%, 10.5%, and 19.0%) and various flap configurations. By further changing the height and the thickness of the Gurney flaps, it was found that the height of the Gurney flaps is a very important parameter, whereas the thickness parameter has little influence, and the maximum lift coefficient of the airfoil with flaps is increased by 8.47% to 13.50% under low turbulent inflow condition.

There is one paper dealing with tip loss correction study. Tip loss correction is important for predicting the aerodynamic performance of a wind turbine, and modelling the tip loss correction is essential in wind turbine aerodynamics. Zhong et al. [10] presented a tip loss correction study for actuator disc/Navier–Stokes simulations with the newly developed tip loss correction model in [11]. The study was conducted to simulate the flow past the experimental National Renewable Energy Laboratory (NREL) Phase VI wind turbine [12] and the virtual NREL 5 MW wind turbine [6]. Three different implementations of the widely used Prandtl tip loss function [13] are discussed and evaluated, together with the new tip loss correction in [10]. It was found that the performance of three different implementations [14–16] is roughly consistent with the standard Glauert correction employed in the blade element momentum theory, but they all tend to make the blade tip loads over-predicted, and the new tip loss correction shows superior performances in various flow conditions.

There is one paper dealing with the development of flow-structure-acoustics framework for predicting and controlling the noise emission from a wind turbine under wind shear and yaw [17]. A wind turbine operating under wind shear and in yaw produces periodic changes of blade loading, which intensifies the amplitude modulation (AM) of the generated noise, and thus can give more annoyance to the people living nearby. In this study, the noise emission from a wind turbine under wind shear and yaw is modelled with an advanced fluid-structure-acoustics framework, and then controlled with a pitch control strategy. The numerical tool used in this study is the coupled Navier–Stokes/Actuator Line model EllipSys3D/AL [18], structure model FLEX5 [19], and noise prediction model (Brooks, Pope, and Marcolini: BPM) [20] framework. Simulations and tests were made for the NM80 wind turbine [21] equipped with three blades made by LM Wind Power. The coupled code was first validated against field load measurements under wind shear and yaw, and a fairly good agreement was obtained. The coupled code was then used to study the noise source control of the turbine under wind shear and yaw.

There is one paper dealing with a study of orthopter-type vertical axis wind turbine (O-VAWT) concept [22]. The study by Wijayanto et al. [23] investigated the effects of horizontal shear flow on the power performance characteristics of an O-VAWT by performing wind tunnel experiments and computational fluid dynamics (CFD) simulations. A uniform flow and two types of shear flow (advancing side faster shear flow (ASF-SF) and retreating side faster shear flow (RSF-SF)) were employed as the approaching flow to the O-VAWT. The ASF-SF had a higher velocity on the advancing side of the rotor. The RSF-SF had a higher velocity on the retreating side of the rotor. It was found that the location where ASF-SFs with high shear strength dominantly occur is ideal for installing the O-VAWT.

3. Future Research Need

Although this Special Issue has been closed, more research in wind turbine aerodynamics is expected, as the goal of wind energy research is to help the technological development of new, environmentally friendly, and cost-effective large wind turbines and wind farms.

Funding: The special issue was funded by the key programs of the Ministry of Science and Technology, grant number 2019YFE0192600 (Research on Key Technologies of Low Noise Wind Turbine).

Acknowledgments: This Special Issue would not be possible without the contributions of various talented authors, professional reviewers, and the dedicated editorial team of Applied Sciences. Congratulations to all the authors. I would like to take this opportunity to record my sincere gratefulness to all the reviewers. Finally, I place my gratitude to the editorial team of Applied Sciences, and special thanks to Nicole Lian, Assistant Managing Editor from MDPI Branch Office, Beijing.

Conflicts of Interest: The author declares no conflict of interest.

References

1. World's Biggest Wind Turbine Shows the Disproportionate Power of Scale. Available online: https://newatlas.com/energy/worlds-biggest-wind-turbine-mingyang/ (accessed on 22 August 2021).
2. Shen, W.Z.; Sørensen, J.N. Special Issue on Aerodynamics of Offshore Wind Energy Systems and Wakes. *Renew. Energy* **2014**, *70*, 1–2. [CrossRef]
3. Shen, W.Z. Special Issue on Wind Turbine Aerodynamics. *Appl. Sci.* **2019**, *9*, 1725. [CrossRef]
4. Sun, Z.; Zhu, W.; Shen, W.; Tao, Q.; Cao, J.; Li, X. Numerical Simulations of Novel Conning Designs for Future Super-Large Wind Turbines. *Appl. Sci.* **2021**, *11*, 147. [CrossRef]
5. Zahle, F.; Bak, C.; Sørensen, N.N.; Guntur, S.; Troldborg, N. Comprehensive Aerodynamic Analysis of a 10 MW Wind Turbine Rotor Using 3D CFD. In Proceedings of the 32nd ASME Wind Energy Symposium, National Harbor, MD, USA, 13–17 January 2014.
6. Jonkman, J.; Butterfield, S.; Musial, W.; Scott, G. *Definition of a 5-MW Reference Wind Turbine for Offshore System Development*; Technical Report NREL/TP-500-38060; National Renewable Energy Laboratory: Golden, CO, USA, 2009.
7. Wang, Q.; Liao, K.; Ma, Q. The Influence of Tilt Angle on the Aerodynami20Performance of a Wind Turbine. *Appl. Sci.* **2018**, *10*, 5380. [CrossRef]
8. Yang, J.; Yang, H.; Zhu, W.; Li, N.; Yuan, Y. Experimental Study on Aerodynamic Characteristics of a Gurney Flap on a Wind Turbine Airfoil under High Turbulent Flow Condition. *Appl. Sci.* **2020**, *10*, 7258. [CrossRef]
9. Sun, Z.; Sessarego, M.; Chen, J.; Shen, W.Z. Design of the of Wind China 5 MW Wind Turbine Rotor. *Energies* **2017**, *10*, 777. [CrossRef]
10. Zhong, W.; Wang, T.G.; Zhu, W.J.; Shen, W.Z. Evaluation of Tip Loss Corrections to AD/NS Simulations of Wind Turbine Aerodynamic Performance. *Appl. Sci.* **2019**, *9*, 4919. [CrossRef]
11. Zhong, W.; Shen, W.; Wang, T.; Li, Y. A tip loss correction model for wind turbine aerodynamic performance prediction. *Renew. Energy* **2020**, *147*, 223–238. [CrossRef]
12. Hand, M.M.; Simms, D.A.; Fingersh, L.J.; Jager, D.W.; Cotrell, J.R.; Schreck, S.; Larwood, S.M. *Unsteady Aerodynamics Experiment Phase Vi: Wind Tunnel Test Configurations and Available Data Campaigns*; NREL/TP-500-29955; National Renewable Energy Laboratory: Golden, CO, USA, 2001.
13. Glauert, H. Airplane propellers. In *Aerodynamic Theory*; Durand, W.F., Ed.; Dover: New York, NY, USA, 1963; pp. 169–360.
14. Sørensen, J.N.; Kock, C.W. A model for unsteady rotor aerodynamics. *J. Wind. Eng. Ind. Aerodyn.* **1995**, *58*, 259–275. [CrossRef]
15. Mikkelsen, R.; Sørensen, J.N.; Shen, W.Z. Modelling and analysis of the flow field around a coned rotor. *Wind. Energy* **2001**, *4*, 121–135. [CrossRef]
16. Shen, W.Z.; Sørensen, J.N.; Mikkelsen, R. Tip Loss Correction for Actuator/Navier–Stokes Computations. *J. Sol. Energy Eng.* **2005**, *127*, 209–213. [CrossRef]
17. Zhou, M.; Sessarego, M.; Yang, H.; Shen, W.Z. Development of an Advanced Fluid-Structure-Acoustics Framework for Pre-dicting and Controlling the Noise Emission from a Wind Turbine under Wind Shear and Yaw. *Appl. Sci.* **2020**, *10*, 7610. [CrossRef]
18. Shen, W.Z.; Zhu, W.J.; Sørensen, J.N. Actuator line/Navier-Stokes computations for the MEXICO rotor: Comparison with detailed measurements. *Wind. Energy* **2012**, *15*, 811–825. [CrossRef]
19. Øye, S. FLEX Simulation of wind turbine dynamics. In *State of the Art of Aeroelastic Codes for Wind Turbine Calculations*; Pedersen, M.B., Ed.; The Technical University of Denmark: Lyngby, Denmark, 1996.
20. Zhu, W.J.; Heilskov, N.; Shen, W.Z.; Sørensen, J.N. Modeling of Aerodynamically Generated Noise from Wind Turbines. *J. Sol. Energy Eng.* **2005**, *127*, 517–528. [CrossRef]
21. Madsen, H.A.; Sørensen, N.N.; Bak, C.; Troldborg, N.; Pirrung, G. Measured aerodynamic forces on a full scale 2MW turbine in comparison with EllipSys3D and HAWC2 simulations. *J. Phys. Conf. Ser.* **2018**, *1037*, 022011. [CrossRef]
22. ElKhoury, M.; Kiwata, T.; Nagao, K.; Kono, T.; Elhajj, F. Wind tunnel experiments and Delayed Detached Eddy Simulation of a three-bladed micro vertical axis wind turbine. *Renew. Energy* **2018**, *129*, 63–74. [CrossRef]
23. Wijayanto, R.P.; Kono, T.; Kiwata, T. Performance Characteristics of an Orthopter-Type Vertical Axis Wind Turbine in Shear Flows. *Appl. Sci.* **2020**, *10*, 1778. [CrossRef]

Article

Numerical Simulations of Novel Conning Designs for Future Super-Large Wind Turbines

Zhenye Sun [1], Weijun Zhu [1,*], Wenzhong Shen [2], Qiuhan Tao [1], Jiufa Cao [1] and Xiaochuan Li [1]

[1] School of Electrical, Energy and Power Engineering, Yangzhou University, Yangzhou 225127, China; zhenye_sun@yzu.edu.cn (Z.S.); MZ120190787@yzu.edu.cn (Q.T.); jfcao@yzu.edu.cn (J.C.); xcli@yzu.edu.cn (X.L.)

[2] Department of Wind Energy, Technical University of Denmark, 2800 Lyngby, Denmark; wzsh@dtu.dk

* Correspondence: wjzhu@yzu.edu.cn; Tel.: +86-139-2160-2334

Abstract: In order to develop super-large wind turbines, new concepts, such as downwind load-alignment, are required. Additionally, segmented blade concepts are under investigation. As a simple example, the coned rotor needs be investigated. In this paper, different conning configurations, including special cones with three segments, are simulated and analyzed based on the DTU-10 MW reference rotor. It was found that the different force distributions of upwind and downwind coned configurations agreed well with the distributions of angle of attack, which were affected by the blade tip position and the cone angle. With the upstream coning of the blade tip, the blade sections suffered from stronger axial induction and a lower angle of attack. The downstream coning of the blade tip led to reverse variations. The cone angle determined the velocity and force projecting process from the axial to the normal direction, which also influenced the angle of attack and force, provided that correct inflow velocity decomposition occurred.

Keywords: coned rotor; aerodynamics; wind turbine; computational fluid dynamics

check for
updates

Citation: Sun, Z.; Zhu, W.; Shen, W.; Tao, Q.; Cao, J.; Li, X. Numerical Simulations of Novel Conning Designs for Future Super-Large Wind Turbines. *Appl. Sci.* **2021**, *11*, 147. https://dx.doi.org/10.3390/app11010147

Received: 3 December 2020
Accepted: 22 December 2020
Published: 25 December 2020

Publisher's Note: MDPI stays neutral with regard to jurisdictional claims in published maps and institutional affiliations.

1. Introduction

Wind turbines are increasing in size and rated power in order to meet the requirement of wind energy development and further reduce the cost of energy (COE). Commercially available wind turbines are reaching 15 MW and are expected to achieve a power levels of 20 MW with even larger rotor diameters. As the blade mass increases subcubically with the blade length [1], the mass per blade would surpass 75,000 kg for a 20 MW wind turbine [2], which will give rise to difficulties in the design and construction of such systems. Adopting carbon fiber laminates in the major load-carrying region, such as the cap, can reduce the blade mass. Smart blades with advanced control strategies, together with add-ons, such as moving trailing edge flaps, can reduce the load and cost [3]. However, the utilization of advanced materials and the smart control techniques is constrained by cost [2]. Blade structure optimization [4–6] can also reduce blade mass. An optimization study on a 5 MW wind turbine rotor [6] found that the blade-tower clearance impedes the further reduction of blade mass, which implies the importance of coned, tilted, and prebending rotors. These ideas are not new, and have already been commercially applied. The coned rotor can reduce the static and dynamic loads [7], which will greatly reduce the blade weight and cost. Prebending blades are manufactured with their stacking lines flexed toward the wind. Compared with coned and tilted rotors, prebending blades can be mounted on the nacelle without the need to modify the design of the latter.

Based on the ultralight, load-aligned rotor concept, a downwind design by Loth et al. [2,8] was proposed to orient the resultant force of blades along the span-wise direction. Their blades are mostly under a tensional force and suffer fewer bending moments than traditional blades. Additionally, downwind rotors have a larger tower clearance. Therefore, this load-aligned downwind design will get rid of the rotor-tower clearance constraint and

5

make the full use of the material strength, which allows more flexible and lighter blades to be manufactured. It was found that a two-bladed rotor following this concept leads to a mass saving around 27%, based on a 13.2 MW reference rotor [9,10]. Qin et al. [11] upscaled the load-aligned design from 13.2 MW to 25 MW. Additionally, segmented blades and outboard pitching ideas were discussed as a means of overcoming the increased edge-wise loads of load-alignment. Wanke et al. [12] compared a 2.1 MW three-bladed upwind turbine with the downwind counterparts. It was concluded that downwind configurations have no clear advantage over the original upwind design. This conclusion was made on the condition that downwind rotors are not specially redesigned for downwind conditions; redesigns may yield different results. Bortolotti et al. [13] compared 10 MW upwind and downwind three-bladed rotors, with and without active cone control. They found that downwind designs, despite having reduced cantilever loadings, did not show obvious advantages over upwind designs. Ning and Petch [14] published an integrated design of 5–7 MW downwind turbines and compared it with its upwind counterparts. It was found that 25–30% of the rotor mass could be reduced at Class III wind sites. The overall cost of energy was reduced by only 1–2%, because the benefits of a reduced rotor mass are offset by a larger tower mass, required to maintain the overhang of the mass center in downwind configurations. They noted the potential to reduce the cost of energy by using downwind rotors, but also acknowledged that more studies are needed. In short, discussions on the downwind and load-alignment concepts are ongoing, and the shift to downwind designs will require more studies.

Inspired by the above studies on load-aligned or downwind concepts, the present paper puts forward a conceptual design, as shown in Figure 1. This concept actively cones the blade tip rather than the whole blade. When wind velocity increases from cut-in speed to rated speed, the blade tip cones further downwind to make the outer part of blade actively load-aligned. However, the inner part of blade is prebended to a fixed load-aligned shape and is fixed to the hub. The maximum thrust of the rotor normally appears near the rated power condition. This new concept can reduce the blade root flapwise bending moments with the alliance of a fixed load-aligned part and an actively load-aligned part. When the wind speed approaches the cut-off speed, the thrust force gradually decreases. Meanwhile, the rotor has a constant rotational speed and a constant centrifugal force, so that the blade tip can cone upwind slightly to meet the new load-aligned condition. Under extreme wind conditions such as typhoons, the blade tip will fold and pitch to a feathered state. This active tip-conning process consumes less energy than the original load-aligned concept [9,10] which cones the whole blade, and may consume a non-negligible amount of power [13]. Additionally, the new concept has a mass center closer to the tower which introduces smaller tower base moments than the original load-aligned concept; this is especially beneficial under very strong winds. Last but not least, the downwind concepts can extend the cut-off wind speed to a larger value (for example 30 m/s), as they have a larger tower-blade clearance compared to the upwind configuration.

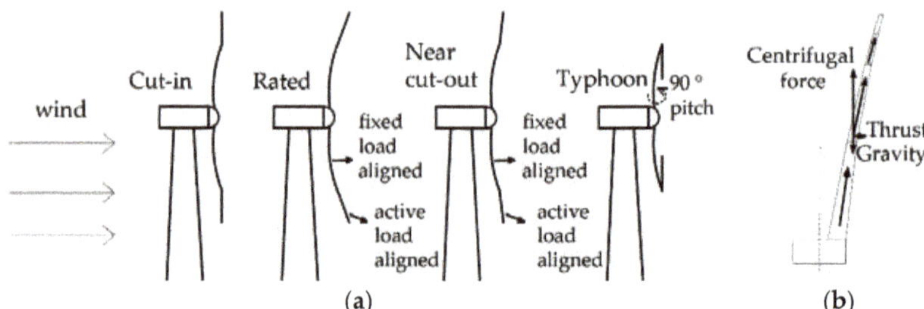

Figure 1. New concept of a combination of a fixed and an active load-alignment: (**a**) blade shape under different wind conditions; (**b**) sketch of load-alignment.

The present paper mainly focuses on the aerodynamic aspects related to these designs. In [9–14], although different simulation tools were utilized, the aerodynamic computations were all based on Blade Element Momentum (BEM) theory. However, classical BEM theory is not suitable for coned rotors, especially with a large cone angle. Mikkelsen et al. [15] applied the traditional BEM method to a coned rotor. It was found that obvious errors appeared, even if a proper decomposition of the inflow velocity on coned blades was made. The inapplicability of the classical BEM method was also noted by Madsen et al. [16] and Crawford et al. [7,17]. Crawford corrected the BEM method by applying a vortex method as well as the proper decomposition of the inflow velocity in the rotor plan [17]. The conclusions in the above studies [9–14] contain strong uncertainties due to the application of classical BEM to coned rotors. So, it is of vital importance to accurately compare the aerodynamic characteristics of these designs, which is the foundation of all of these concepts. Nevertheless, aerodynamic research on coned, tilted and prebended rotors is very limited. Notably, computational fluid dynamics (CFD) methods with three dimensional (3D) body-fitted meshes are scarce. Actuator disc (AD) CFD methods are one of the commonly used numerical methodologies. Madsen and Rasmussen [18] compared four downwind rotors utilizing the AD CFD method, and found that the span-wise axial induction distributions and power coefficient were obviously influenced by the out of plane bending. With the help AD CFD, Mikkelsen et al. [15] also found a similar influence of coning on inducted velocities. It was also found that the upwind conning had a 2–3% higher power coefficient than the downwind configuration. However, AD-based methods are inherently coupled with BEM, which has some limitations. Winglets can be seen as partially coned blades. Farhan et al. [19] utilized a 3D CFD method to analyze the effect of winglets, observing that their influence extended to 30% of the radial sections. The phenomenon whereby the uncurved part of the blade was influenced by the deformed part was also observed in the study [18]. The Vortex Method (VM) can also be adopted to analyze such rotors. Chattot [20] utilized the VM method to investigate the influence of different blade tip configurations such as sweep, bending and winglets, and found that the whole blade was influenced by the curved part. Additionally, it was found that upwind prebending yielded increased power compared to the downwind configuration, which agreed with research presented in [15,18,19]. Further study is needed to understand the nonlinear behavior related to blade bending, as noted by Chattot [20]. Shen et al. [21] utilized the VM method to optimize rotor blades and found that the bended blade tip had an aerodynamic influence on the whole blade, and that this could not be accounted for using the traditional BEM method. Lastly, wind tunnel experiments could be conducted to explore these rotor concepts [22–24]. Due to their complexity, experiments to date have only investigated overall performance, such as thrust and torque, rather than span-wise force distribution. Therefore, dedicated CFD simulations are indispensable, especially on full-scale wind turbine rotors with 3D body-fitted meshes. Prebending has a continuously changing slope or cone angle, so it is hard to quantify its effects. Conning is the basis of prebending, and conning designs are suitable for parametric studies. In a previously published paper [25], the authors simulated different up/downwind conning and presented a preliminary aerodynamic analysis. The present paper will analyze the aerodynamic performance of different conning effects, such as inflow velocity decomposition and angle of attack analysis. Additionally, this paper will cover novel cones with three segments, which is a simplification and standardization of the new concepts shown in Figure 1.

The paper is organized as follows. In Section 2, the configurations of the cones and the employed numerical methods are presented. Results and discussions are given in Section 3. Finally, conclusions are drawn in Section 4.

2. Modeling and Methods

2.1. Modelling of Different Cone Configurations

In order to analyze the aerodynamic performance of the load-aligned concepts, different conning configurations were designed, as shown in Figures 2 and 3. These configurations are transformed from the DTU-10-MW Reference Wind Turbine (RWT) [26,27] rotor, which is referred to as the baseline rotor, according to Equations (1) and (2). To focus on the effects of coning, a DTU-10-MW RWT without cone, tilt, prebend, nacelle or hub was used. At radial position r, coned configurations had their blade stacking lines translated out of the rotor plane with a displacement of Z_{cone}.

$$Z_{cone} = \begin{cases} 0, & r \leq T_{trans}R \\ (r/R - T_{trans})R/C_{cone}, & r > T_{trans}R \end{cases} \tag{1}$$

where R is the rotor radius, T_{trans} is the relative radial position where cone starts, and C_{cone} controls the slope of the stacking line. As shown in Figure 2a, $T_{trans} = 5/R$ means cone starting at 5 m, and $T_{trans} = 1/3$ means cone starting at $R/3$. The cone angles are controlled by C_{cone}. When $C_{cone} = \pm 4, \pm 8$, the rotors have cone angles of $\pm 14.0362°$, $\pm 7.1250°$, respectively. A larger $|C_{cone}|$ produces a smaller cone angle, and a positive Z_{cone} makes a downwind cone. As shown in Figure 2b, several special coning configurations are shown, which are further coned at $2R/3$. These special cone designs have an out of plane displacement of Z_{cone}, as defined by Equation (2).

$$Z_{cone} = \begin{cases} 0, & r \leq T_{trans}R \\ (r/R - T_{trans})R/C_{cone}, & T_{trans}R < r < 2R/3 \\ (2/3 - T_{trans})R/C_{cone}, & r \geq 2R/3, \text{ for S0} \\ (4/3 - T_{trans} - r/R)R/C_{cone}, & r \geq 2R/3, \text{ for S1} \\ (2r/R - T_{trans} - 2/3)R/C_{cone}, & r \geq 2R/3, \text{ for S2} \end{cases} \tag{2}$$

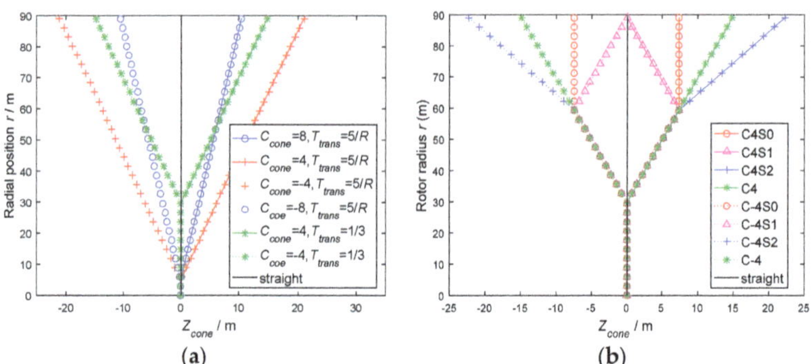

Figure 2. Conning configurations: (**a**) $T_{trans} = 5/R$ with $C_{cone} = \pm 4, \pm 8$ and $T_{trans} = 1/3$ with $C_{cone} = \pm 4$; (**b**) special configurations abbreviated as C4S0, C4S1, C4S2, C4, C-4S0, C-4S1, C-4 S2 and C-4.

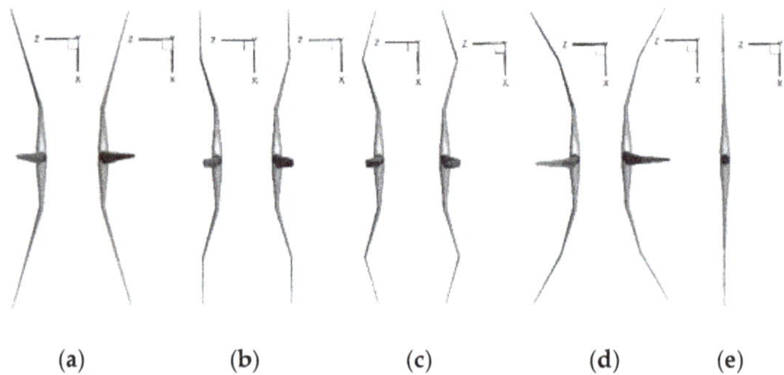

Figure 3. Down/upwind coned configurations with $C_{cone} = \pm 4$ and $T_{trans} = 1/3$: (**a**) C4 and C-4; (**b**) C4S0 and C-4S0; (**c**) C4S1 and C-4S1; (**d**) C4S2 and C-4S2; (**e**) straight baseline.

In the rest part of the paper, a name abbreviation rule is applied for $T_{trans} = 1/3$ configurations. $C_{cone} = \pm4$ and $T_{trans} = 1/3$ are named as C4 and C-4, respectively. Symbols S0, S1 and S2 are used to discriminate among the configurations at $r > 2R/3$. For example, the case of $C_{cone} = \pm4$ and $T_{trans} = 1/3$ followed by S2 is abbreviated as C4S2 and C-4S2, which have their blade tips farthest from the rotor plane. The symbol of S1 represents a reverse blade tip cone, such that C4S1 has its blade tip pointing to the upwind direction of a downwind cone at $r < 2R/3$. Lastly, the symbol S0 represents a zero cone angle at $r > 2R/3$. In short, C4 stands for downwind and C-4 for upwind, and S0, S1 and S2 represent blade tip configurations. For the sake of aerodynamic comparisons, all the configurations have the same projected areas and the same distribution of airfoil thickness, chord and twist. The shapes of different configurations are depicted in Figure 3, with the wind flow from the negative Z to the positive Z.

2.2. Mesh Structure and CFD Method

The baseline rotor has been studied elsewhere [26–28]; past studies provided references for the mesh configuration applied here. The mesh employed a commonly used O-O configuration with the surface mesh on one blade containing 256 points in the airfoil circumferential direction and 128 points in the span-wise direction. The volume mesh was expanded from the surface mesh to the far-field boundary (approximately 17R away) with 128 cells along the normal direction. To meet the computational requirement of Y + <2, the first cell height was 2×10^{-6} m. Finally, the grid was constructed with 432 blocks which contained 14.16 million structural cells in total. A similar mesh configuration is accurate enough to simulate the aerodynamic performance of the DTU 10 MW RWT rotor [26,28] which can be found on the DTU 10 MW RWT project website [27]. Such mesh settings were used for all the coned configurations in the present paper. The blade surface mesh of C4S1 is shown in Figure 4a, and the mesh distributions on two cross-sections are illustrated in Figure 4b,c.

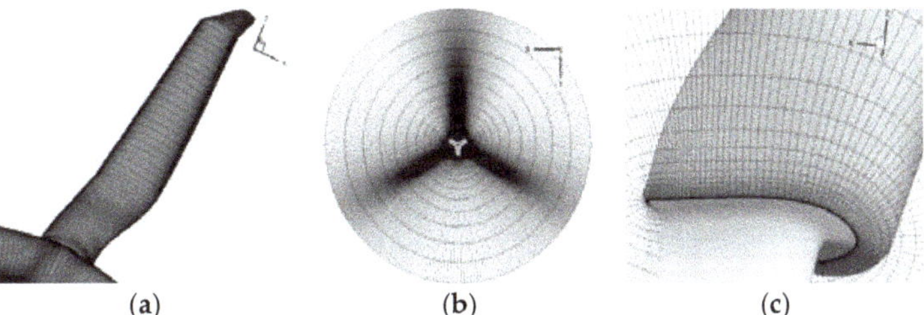

(a)   (b)   (c)

Figure 4. Mesh around the blades: (**a**) blade surface mesh; (**b**) mesh on a section away from the near-blade blocks; (**c**) mesh on an airfoil cross-section in the near-blade region.

The flow state is treated as incompressible, and the turbulence flow is fully developed. The flow-field was solved by the Reynolds-Averaged Navier-Stokes (RANS) equations with the $k - \omega$ SST turbulence model [29]. The SIMPLE algorithm was utilized to couple the pressure and velocity equations. EllipSys3D, developed by the Technical University of Denmark and widely validated over the past 20 years, was used as the CFD solver. Detailed descriptions of the solver can be found in [26,30]. Additionally, detailed boundary condition descriptions and baseline rotor validation can be found in a previous publication [25], where the same numerical methods were adopted. At a wind speed of 12 m/s, few force differences appeared between steady and unsteady simulations in the root region ($r < R/3$) where flow separations and 3D rotational augmentations were expected [28]. The forces along the outer part of blade remained identical. In this paper, steady CFD simulations were performed to investigate the influence of coning at a wind speed of 9 m/s, i.e., at which the unsteady effects were negligible. The operational parameters of the baseline rotor, listed in Table 1, were applied for all the coned configurations.

Table 1. The operational parameters.

Wind Speed (m/s)	Pitch (Degree)	Rotational Speed (RPM)
9.000	0.000	7.229

3. Results and Discussions

In this section, the aerodynamic performance of the coned DTU 10 MW rotor, coning at a blade position near the root ($T_{trans} = 5/R$), is presented. In order to explain the physics behind the coned rotor, the concept of angle of attack (AOA) on the rotor blade sections was extended to include coned rotors. The results are discussed through the concept of angle of attack.

3.1. Four Configurations of Coning Near the Root: $T_{trans} = 5/R$ and $C_{cone} = \pm4, \pm8$

3.1.1. Force Performances

Firstly, the overall aerodynamic performance of the configurations presented in Figure 2a are compared. The two configurations of $C_{cone} = \pm 4$ and $T_{trans} = 1/3$ also appear in Figure 2b, where they are abbreviated as C4S2 and C-4S2. These configurations are not discussed here, and will be explored in Section 3.2. In Table 2, the thrust T and torque Q of the other four configurations in Figure 2a are listed. As high torque and low thrust are beneficial, the torque-to-thrust ratio (QT) was used to compare different conning configurations. The relative variations of these parameters are denoted as δT, δQ and δQT; for example, δT means

$$\delta = \frac{|T|_{cone} - |T|_{straight}}{|T|_{straight}} \times 100\% \tag{3}$$

Table 2. Thrust and torque of different configurations ($T_{trans} = 5/R$).

	Straight	$C_{cone} = 8$	$C_{cone} = -8$	$C_{cone} = 4$	$C_{cone} = -4$
T(KN)	1046.06	1057.24	1025.85	1060.77	998.63
δT	0.00%	1.07%	−1.93%	1.41%	−4.53%
Q(KNm)	7283.11	7254.96	7268.08	7195.15	7205.48
δQ	0.00%	−0.39%	−0.21%	−1.21%	−1.07%
QT(m)	6.96	6.86	7.08	6.78	7.22
δQT	0.00%	−1.44%	1.76%	−2.58%	3.63%

The most upwind-coned configuration $C_{cone} = -4$ gives the lowest thrust, i.e., 4.53% lower than that of the baseline without coning. Although $C_{cone} = -4$ reduces the torque by 1.07% compared with the baseline, it has the highest QT due to the obvious decline of thrust. The downwind counterpart $C_{cone} = 4$ produces the highest T, lowest Q, and lowest QT, which is unfavorable. Among the pair of $C_{cone} = \pm8$, the upwind configuration also has a smaller T, a higher Q and a higher QT than the downwind counterpart.

To understand the overall performance differences shown in Table 2, the tangential and axial force per unit span length is shown in Figure 5. The axial force Fz is parallel to the rotor axis, and the tangential force Ft is perpendicular to Fz. The aforementioned forces are the summation of all the three blades. Near the blade tip, the upwind configurations had a larger Ft and Fz than the downwind counterparts. For example, the Ft and Fz curves of $C_{cone} = -4$ were higher than those of $C_{cone} = 4$. Towards the blade root, the situation reversed, i.e., the upwind configurations had a lower Ft and Fz. For the distribution of Ft, the upwind and downwind counterparts had an almost reversed Ft distribution of the baseline rotor with straight blades. The upwind configurations had a higher Ft near the blade tip, which is more beneficial to an increase of torque. As a result, the torque of $C_{cone} = -4$ and −8 was slightly higher than $C_{cone} = 4$ and 8, as listed in Table 2. However, all four coned configurations had a smaller torque than the baseline. For the distribution of *Fz*, the baseline did not lie in the middle of an up/downwind pair, especially toward the blade tip. Although upwind lines gradually surpassed their downwind counterparts and approached the baseline near the blade tip, they barely went above the baseline. The largest upwind cone $C_{cone} = -4$ showed the overall lowest *Fz*, which was consistent with the results shown in Table 2. It is difficult to understand why the force distribution behaved like this, so more analyses are needed.

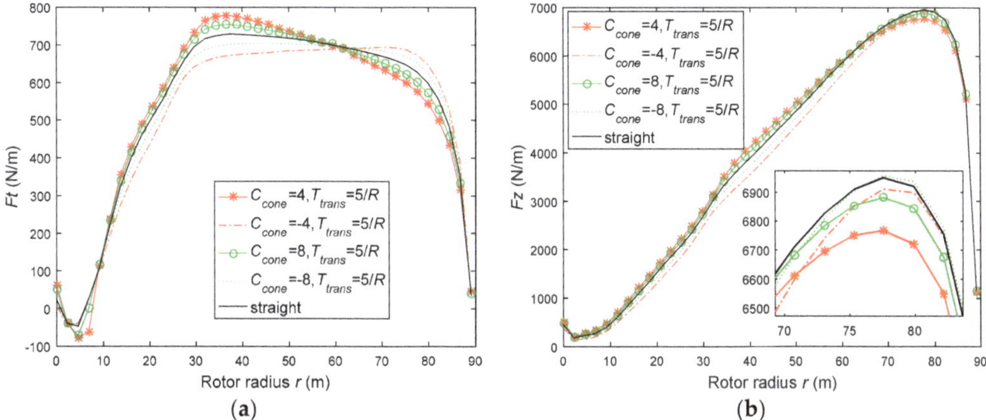

Figure 5. Force distributions along radial direction: (**a**) tangential force; (**b**) axial force.

3.1.2. Flow Field Analysis

Analyzing the flow field around a wind turbine, such as inflow velocity and angle of attack (AOA), may help to understand the force distributions mentioned above. The inflow velocity decomposition for the upwind and downwind coned rotors is illustrated in Figure 6. This decomposition was made in the *YZ* plane, which contained the rotor axis and the pitch axis of a blade. The unit vectors *z*, *r*, *s* and *n* were along the axial, radial, span-wise and normal directions, respectively. At a far upstream position, the axial velocity was V_0, the normal velocity component was $V_0\cos\beta$, and the radial velocity was zero. Towards the rotor, the axial and normal velocity decreased and the radial velocity increased. Arriving at the rotor, the axial and normal velocity was reduced to V_0-Wz, $V_0\cos\beta$-Wn, and the radial velocity increased to Vr. Here, Wz is the axial induction velocity at the aerodynamic center (AC) and Wn is the normal induction velocity at AC.

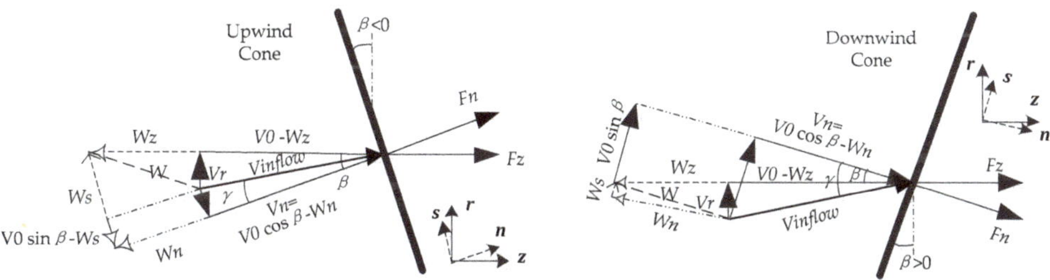

Figure 6. Inflow velocity decomposition for an upwind and a downwind coned rotor.

Due to the presence of *Vr*, the resultant velocity *Vinflow* in the *YZ* plane was not horizontal and did not equal to V_0-Wz, as shown in Figure 6. The velocity decomposition was different for the upwind and downwind coned configurations. For an upwind cone configuration, projecting the resultant velocity *Vinflow* to the normal component was different to that on a downwind counterpart, as the projection angle γ was different. However, the radial velocity component is omitted in the traditional BEM method because there is no equation to describe the radial flow. Then, it was assumed that the inflow velocity *Vinflow* would be along the axial direction and would be equal to V_0-Wz. When projecting the inflow velocity *Vinflow* to the normal component, the projection angle γ equaled the cone angle β. Thus, the upwind and downwind pair had the same value of $\cos\beta$, *Fn* and *Fz*. In short, the traditional BEM methods cannot distinguish the upwind and downwind coning, which is clearly in contrast with reality.

Another way to analyze the flow is to use the concept of angle of attack, which is conducted on the planes perpendicular to the blade spanwise direction. At present, difficulties or uncertainties remain in the extraction of AOA from 3D CFD simulations or experiments, which is different from that in the 2D situations. There are different kinds of methods to extract AOA from CFD data, which are thoroughly discussed and compared in [31,32]. Most methods predicted similar AOA at midspan, but it should be kept in mind that the results near the blade root and tip varied from one method to another. The Average Azimuthal Technique (AAT) [31–33], proposed by Hansen et al. [33], was used in the present paper. At a given rotor radius, AAT extracts the inflow velocity on two circles which are just upstream and downstream of the rotor, as shown in Figure 7a. The number of points on a circle is 72 in the present study. AAT estimates the velocity at AC by averaging the up- and down- stream data, and then calculates AOA using Equation (4). However, Equation (4) is only applicable for straight blades without cones. A general form of AOA, which is suitable for coned rotor analyses, is Equation (5).

$$\alpha_z = \tan^{-1}\left(\frac{Vz}{Vt}\right) - \theta = \tan^{-1}\left(\frac{V_0 - Wz}{Vt}\right) - \theta \tag{4}$$

$$\alpha_n = \tan^{-1}\left(\frac{Vn}{Vt}\right) - \theta = \tan^{-1}\left(\frac{V_0\cos\beta - Wn}{Vt}\right) - \theta \tag{5}$$

where Vz is the estimated axial velocity at AC, Vn is the estimated normal-wise velocity at AC, Vt is the estimated tangential velocity at AC, and θ is the local pitch angle. When applied to straight blades without a cone, Equation (5) is reduced to Equation (4), because Vn becomes Vz. As most studies available on AOA extraction are for straight blades without coning, Equation (4) is commonly used. To extract AOA, the distance from AC to the up-/down- stream annulus was set to one local chord length C, as shown in Figure 7b, where the axial velocity in the YZ plane is also illustrated.

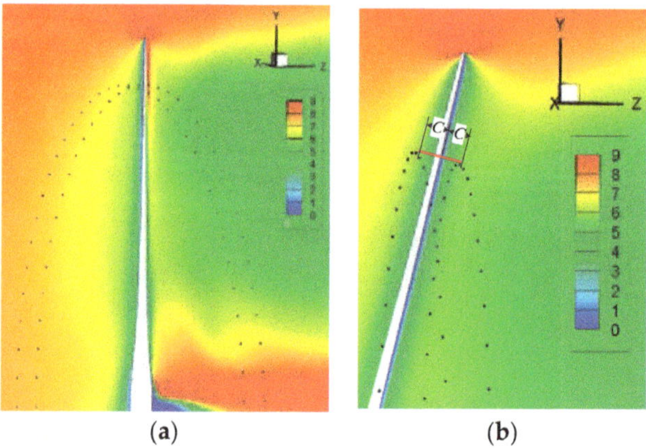

(a)   (b)

Figure 7. Sketch of points used by AAT to extract AOA: (a) straight baseline; (b) coned rotor.

The streamlines in the YZ plane of the two coned configurations are drawn in Figure 8, where the axial velocity contour is also shown. The streamlines of the downwind coned case in Figure 8b were more up-pointing than those of the upwind coned case in Figure 8a. Especially near the blade tip, the upwind cone had a smaller projection angle γ and had streamlines which were more perpendicular to the blade. When projecting the real inflow velocity *Vinflow* in direction n, the upwind coned rotor had a smaller Vn value, even if with the same *Vinflow* value. Therefore, it is more straight forward to analyze α_n, which reveals the inflow condition in the normal-wise Xn plane.

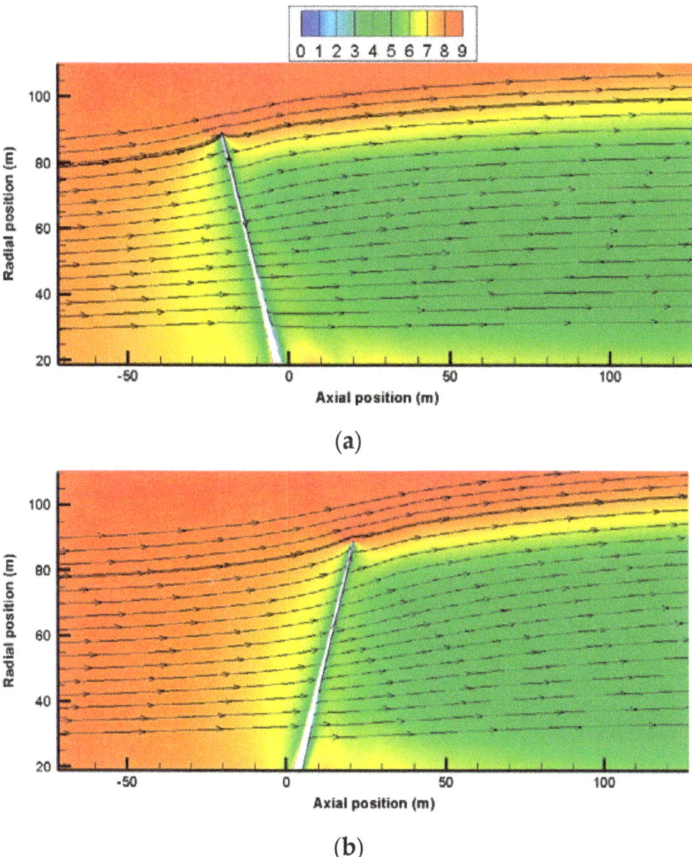

Figure 8. Streamlines in the YZ plane with axial velocity contours for the two coned configurations: (**a**) upwind coned rotor of $T_{trans} = 5/R$, $C_{cone} = -4$; (**b**) downwind coned rotor of $T_{trans} = 5/R$, $C_{cone} = 4$.

To explore the mechanism behind the interesting force distributions shown in Figure 5, the distribution of α_z and α_n are compared in Figure 9. It was found that the α_z of an upwind cone is always smaller than its downwind counterpart, which is not consistent with the force distributions. Interestingly, the α_n curve of the upwind cone intersected with its downwind counterpart. Near the blade root, downwind configurations had a larger α_n than their upwind counterparts, which indicated larger thrust and tangential force. Towards the blade tip, the downwind coning made the α_n gradually decrease below that of upwind cone, which was consistent with the force distribution. Additionally, as shown in Figure 5, there was a phenomenon whereby an up/downwind pair had an almost reversed Ft distribution relative to the baseline, but had a less symmetric distribution of Fz curves, especially towards the tip. A reasonable explanation is given below. The upwind cone had a slightly larger α_n than the straight baseline toward the blade tip, and therefore, also had a larger Fn, which is the normal force parallel to n, as shown in Figure 6. But the force Fz in Figure 5b was along the axial direction, which has a relationship with Fn as follows:

$$Fz \cdot dr = Fn \cdot \cos\beta \cdot dr \tag{6}$$

where the force along the span-wise direction is usually small and thus neglected. Although Fn of the upwind coning was slightly larger than the baseline near the tip, after projecting Fn to Fz, the Fz may have been smaller than the baseline, as shown in Figure 5. It is known that a cone angle always leads to a $\cos\beta$ which is smaller than one; however, there is no projection process for the tangential force Ft. Therefore, the Ft curves of the upwind cone shown in Figure 5 could be higher than the baseline curves, which follow the α_n curves more closely. Lastly, uncertainties still lie in the extraction of AOA, especially near the blade tip [31,32], but it is clear that this provides a view to explain the force distribution shown in Figure 5.

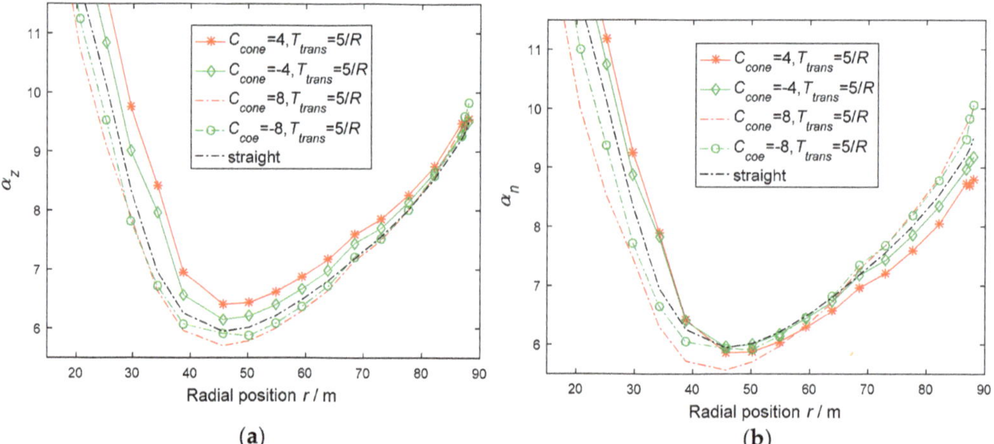

Figure 9. Distributions of angle of attack: (a) α_z; (b) α_n.

To validate the extracted α_n in Figure 9, streamlines and pressure plots around the normal blade section at $r = 77.59$ m are shown in Figure 10. This slice was normal-cutting, which was parallel to the normal-wise Xn plane. It was found that the three airfoil sections were all in an attached flow condition. The α_n of these three configurations could be approximately compared by analyzing the slopes of the streamlines ahead of the leading edge. As the slopes of the streamlines in Figure 10a–c only had minor differences, representative streamlines near the stagnation point were extracted from Figure 10a–c and compared in Figure 10d. It is shown that the upwind coning had a slightly larger α_n than the downwind counterpart, which was consistent with the findings shown Figure 9b; meanwhile, the straight baseline lies in the middle. Additionally, it was clear that the upwind configuration had a lower pressure on the suction-side leading edge than its downwind counterpart, which was in agreement with the larger force of the upwind coning described in Figure 5.

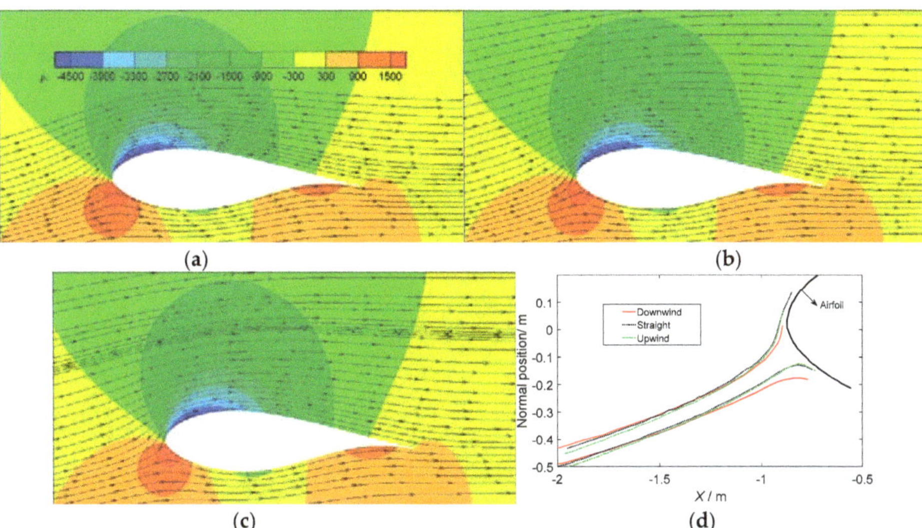

Figure 10. Streamlines and pressure on r = 77.59 m normal-cut plane: (**a**) upwind coning of T_{trans} = 5/R, C_{cone} = −4; (**b**) straight baseline; (**c**) downwind coning of T_{trans} = 5/R, C_{cone} = 4; (**d**) comparisons of streamlines near stagnation points.

3.2. Special Coned Configurations: C4S0, C4S1, C4S2, C4, C-4S0, C-4S1, C-4 S2 and C-4

3.2.1. Overall Force Performance

Configuration C-4S2 gives the lowest thrust among the cases listed in Table 3, i.e., 7.30% lower than the baseline rotor. C-4S2 had the largest blade tip offset among the upwind configurations. Although C-4S2 reduced the torque by 1.96%, it still had the largest torque-to-thrust ratio QT due to the large reduction of thrust. The downwind counterpart C4S2 produced the lowest Q and the lowest QT, which was unfavorable. The upwind configuration surpassed its downwind counterpart, as also revealed in the pairs of C4 and C-4, C4S0 and C-4S0. For these pairs, the upwind coned rotors had a smaller T, a larger Q and a higher QT than their downwind coned counterparts. However, for the pairs C4S1 and C-4S1, the downwind configuration C4S1 had a higher QT. Interestingly, the blade tip of the downwind configuration C4S1 was pointing upwind, which may be the reason why C4S1 had a higher QT. Lastly, it should also be noted that the radial velocity component is omitted in the traditional BEM method. Therefore, the same results will be obtained for an upwind configuration and its downwind counterpart, such as C4S2 and C-4S2, which is clearly in contrast with reality. In Table 3, it may be seen that the thrust discrepancy between C4S2 and C-4S2 reached nearly 8%. The torque discrepancy was approximately 5%. These results reveal that the inaccuracies of traditional BEM methods are not negligible, making the conclusions from [9–14] in Section 1 disputable.

Table 3. Thrust and torque of different configurations.

	Straight	**C4S2**	**C4**	**C4S0**	**C4S1**	**C-4S1**	**C-4S0**	**C-4**	**C-4S2**
T(KN)	1046.06	1051.37	1061.10	1054.99	1032.59	1040.55	1031.22	1005.89	969.73
δT	0%	0.51%	1.44%	0.85%	−1.29%	−0.53%	−1.42%	−3.84%	−7.30%
Q(KNm)	7283.11	7055.26	7197.34	7261.64	7254.33	7250.16	7280.18	7242.09	7140.14
δQ	0%	−3.13%	−1.18%	−0.29%	−0.40%	−0.45%	−0.04%	−0.56%	−1.96%
QT(m)	6.96	6.71	6.78	6.88	7.03	6.97	7.06	7.20	7.36
δQT	0%	−3.62%	−2.58%	−1.14%	0.90%	0.07%	1.40%	3.41%	5.75%

3.2.2. Distributed Force Performances

The axial force Fz and tangential force Ft per unit length are compared in Figure 11. Figure 2b is redrawn in Figure 11a, where the upwind configurations are denoted by the dashed lines and downwind by the dotted lines. Clearly, C4S0 and C-4S0 had the same configuration as the straight baseline when $r > 2R/3$, or in other words, without coning. As a result, the Ft and Fz curves of C4S0, C-4S0 and the straight baseline were very close. In the same spanwise range, C4 and C-4S1 had the same cone angle, as did C-4 and C4S1. Correspondingly, the same cone angle led to close Ft and Fz curves. The discrepancy between the close curves increased towards $r = 2R/3$, because the coning point at $r = 2R/3$ distorted the nearby flow. In short, the same cone angle near the tip will lead to close force distribution.

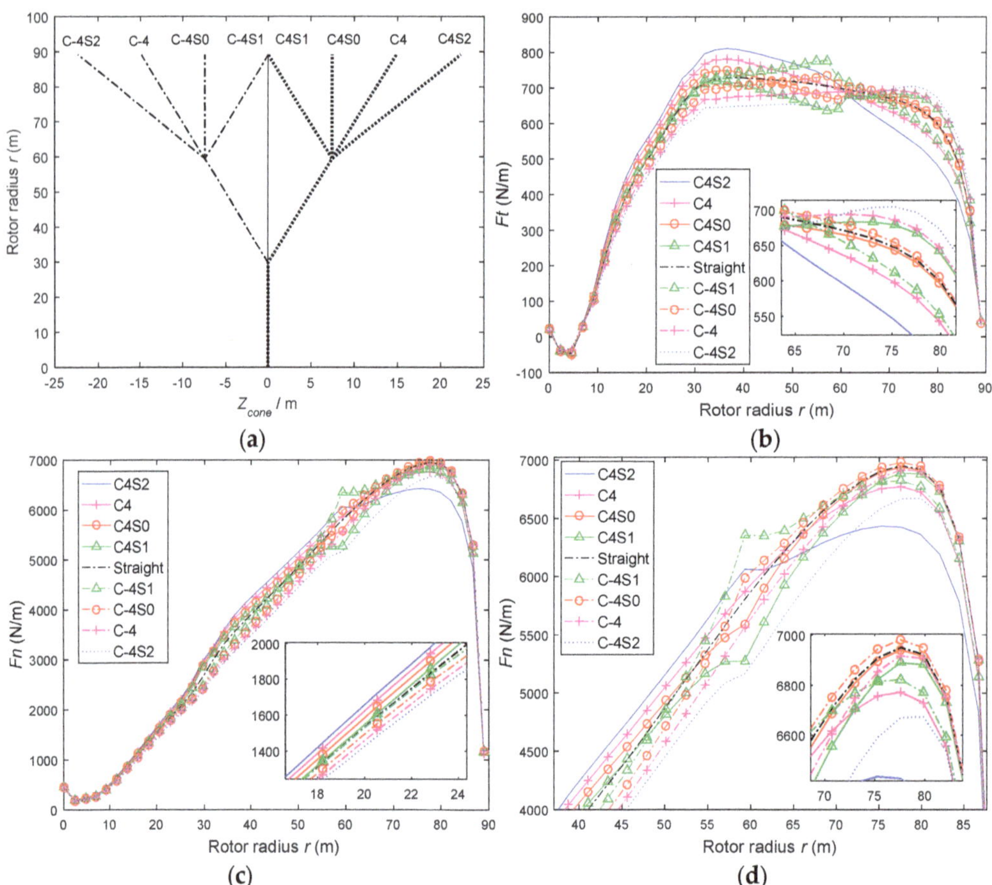

Figure 11. Comparison of special coning: (**a**) redrawn of special cone configurations; (**b**) tangential force; (**c**) axial force; (**d**) zoomed view of axial force.

When $R/3 < r < 2R/3$, the four configurations C4S2, C4, C4S0, C4S1 had the same cone angle as shown in Figure 11a. Additionally, their counterparts C-4S2, C-4, C-4S0, C-4S1 had the same cone angle as well. However, none of the Ft and Fz curves coincided, even if the same cone angles existed, as shown in Figure 11b,c, which indicated that the cone effect in this range was not solely controlled by the cone angle itself. Additionally, traditional BEM methods will predict the same force distribution under the same cone angle, which implies that such an approach is not applicable here. When $r < R/3$, all the cone configurations coincided with the straight baseline. However, only C4S1 and C-4S1 had close force distributions comparable to the straight baseline. The force distribution of $C \pm 4S2$, $C \pm 4$ and $C4 \pm S0$ varied from configuration to configuration. It was found that C4S1 and C-4S1 were totally different cone configurations at $r > R/3$, but that they had the same blade tip position as the straight baseline. This implies that the influence of the coned part on the straight part is mostly determined by the blade tip position. Traditional BEM will predict the same force distribution again, or fail at $r < R/3$, even if all the configurations have a zero cone angle.

There are many interesting phenomena between the curves in Figure 11. Looking closely at group C4S2, C4, C4S0, and C4S1, the Ft and Fz curves (especially Ft lines) are nearly parallel with each other in the range of $R/3 < r < 2R/3$. The case C4S2 had the highest force curves, and C4, C4S0, and C4S1 had successively lower forces. Coincidently, this was consistent with the successively upstream-moving of the tip positions from $Z = 3Z_{tip}$ to $2Z_{tip}$, Z_{tip} and 0 m ($Z_{tip} = 7.4292$ m). If the blade tip is located at a more upstream position, it will cause the blade sections to be further immersed in the wake, which will lead to a stronger axial induction velocity Wz, a smaller axial inflow velocity, and a lower α_z and α_n. The variation of α_z and α_n will be validated later in Section 3.2.3. In short, the upwind transformation of the blade tip is consistent with the successive reduction of the Fz and Ft curves. Focusing on the four counterparts, C-4S2, C-4, C-4S0, and C-4S1, the curves were nearly parallel and successively ascending with the downstream-moving of the blade tips. Transforming the blade tip into a further downstream position, the blade sections immerse less heavily into the tip vortex trace, leading to a smaller Wz and a higher α_n. Additionally, the nearly parallel curves of C4S2, C4, C4S0, and C4S1 had different slopes compared with those of C-4S2, C-4, C-4S0, and C-4S1, and apparently different slopes compared with the baseline rotor.

3.2.3. Flow Field Analysis

To understand the force characteristics presented in Section 3.2.2, further flow field analyses were carried out. Firstly, the streamline and the axial velocity contour in the YZ plane of $C \pm 4S2$, $C \pm 4S0$, and $C \pm 4S1$ are shown in Figure 12. In the range of $R/3 < r < 2R/3$, the downwind cone C4S2 shown in Figure 12a had obviously lower velocity in the near wake region than that of C-4S2 shown in Figure 12b. This revealed that more energy was extracted by C4S2, which is consistent with the higher thrust force in Figure 11. Additionally, the downwind C4S2 had slightly larger wake expansion, which means a larger radial velocity. If the three figures on the left hand side are compared, the wake deficit is weaker and weaker when the blade tip is successively moving upstream from Figure 12a to Figure 12c,e. This means that the energy extracted by the rotor was progressively smaller, which is in agreement with the successive decline in the Ft and Fz curves in the range $R/3 < r < 2R/3$ in Figure 11. If the right hand side figures are compared, the wake deficit becomes stronger when the blade tip transforms downstream. This also confirms the successive increases in the Ft and Fz curves from C-4S2 to C-4S0 and then C-4S1. In the range $r > 2R/3$, the streamlines of C4S1 and C4S2 were the most and the least perpendicular streamlines *w.r.t.* to the blade, respectively, which led to the largest and smallest Fn. But, as shown in Equation (6), the large cone angle β reduces the value of Fz, which causes the Fz of C4S1 to barely surpass the baseline.

Utilizing the AAT AOA-extraction method introduced in Section 3.1.2, the α_z and α_n at different radial positions are compared in Figure 13. In the range of $r > 2R/3$, the α_z and α_n of C4S0 and C-4S0 nearly coincided with the straight baseline, which was consistent with the close Ft and Fz curves, as shown in Figure 11. This was because the three configurations were all without coning. What is more, C4 and C-4S1 had the same cone angle, which led to close α_n distributions. Similarly, C-4 and C4S1 also had close α_n distributions. In the range of $R/3 < r < 2R/3$, the four configurations, C4S2, C4, C4S0 and C4S1, had nearly parallel α_n curves. The α_n line of C4S2 was the highest, and C4, C4S0 and C4S1 had successively lower curves, which matched the successive decreases of the *Ft* and *Fz* curves in Figure 11. As discussed in Section 3.2.2, this was caused by the upstream movement of the tip positions, which made caused the blade sections to be further immersed into the wake, leading to decreases in α_z and α_n. In contrast, focusing on the group C-4S2, C-4, C-4S0, and C-4S1, the α_n curves successively ascended with the downstream movement of the blade tips. In the range of $r < R/3$, C4S1 and C-4S1 had similar α_z and α_n distributions to the straight baseline. This was because the three configurations had the same blade tip position, although distinctly different cones appeared at $r > R/3$. Generally speaking, the α_n distributions matched the force distributions in Figure 11. It is clear that the α_n distributions can reveal the mechanism of force distributions on coned sections, even if the blades are coned into three parts. Additionally, correctly commutating α_n is of vital importance for improving the traditional BEM method, although further discussion of this is beyond the scope of this paper.

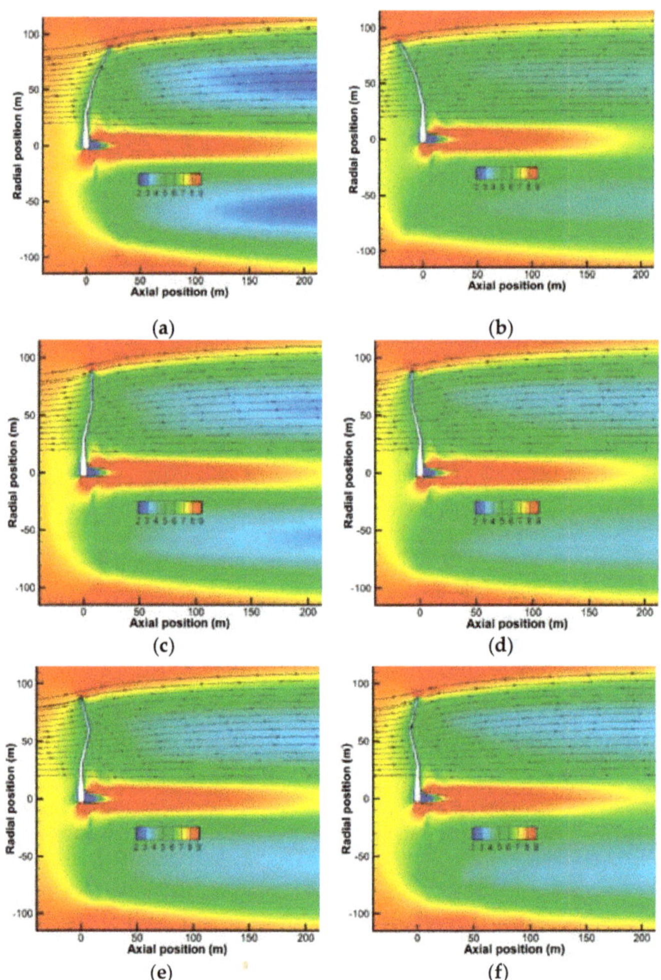

Figure 12. Streamline in the YZ plane with axial velocity contours: (**a**) C4S2; (**b**) C-4S2; (**c**) C4S0; (**d**) C-4S0; (**e**) C4S1; (**f**) C-4S1.

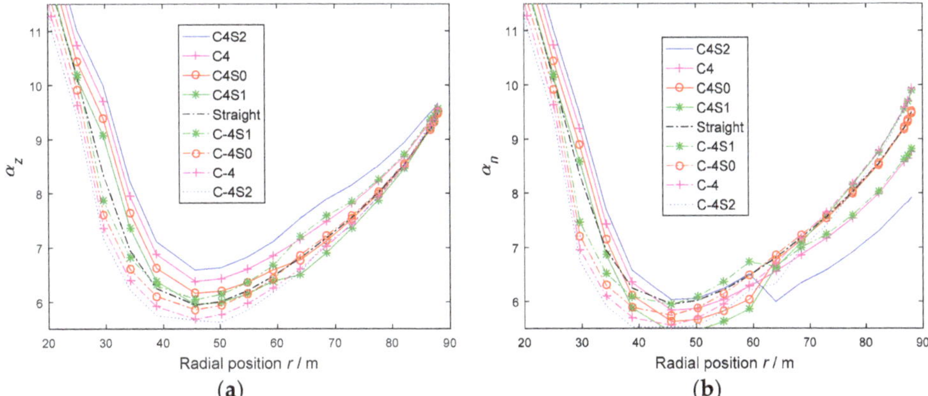

Figure 13. Distributions of angle of attack: (**a**) α_z; (**b**) α_n.

4. Conclusions

In future designs of super-large wind turbines, the question of being upwind or downwind will be an important one for the wind energy industry. The present paper put forward a conceptual design consisting of an actively load-aligned blade tip and a fixed load-aligned blade root. In order to evaluate the advantages and disadvantages of these concepts, it is of vital importance to carefully select appropriate tools. Different conning configurations, including special cones with three segments, were simulated and analyzed based on a 10 MW reference rotor. The results provide knowledge regarding the complex force distributions of these configurations, and could serve to improve the traditional BEM on traditionally or specially coned rotors.

Up- and down- wind coning approaches yield different aerodynamic performance, e.g., in their total integrated loading, distributed force, and flow fields. The force distributions and their differences may be explained by the concept of the angle of attack. It was found that parameters which have the greatest influence on the angle of attack are the position of blade tip and the cone angle. The blade tip position determines the induction velocity contour, and subsequently, the inflow velocity at the blade sections. With the upstream movement of the blade tip, blade sections away from the tip will be immersed more heavily in the tip vortex trace. Then, the blade sections will suffer from stronger axial induction, smaller axial inflow velocity, a lower angle of attack, and consequently, a lower force distribution. The downstream movement of the blade tip has the opposite effect. The cone angle determines the velocity and force projecting process from the axial to the normal direction, which influences the thrust and tangential forces in the normal and axial directions. The correct inflow velocity decomposition, which connects the axial and normal directions, is indispensable. The same relative blade tip position and cone angle will result in the same force; however, applying the same tip position or cone configuration alone does not guarantee the same aerodynamic performance.

The aerodynamic performance discrepancy between an upwind cone and its down-wind counterpart is significant. In the present study, the most upwind and downwind cones had a thrust difference up to 8% and a torque difference of up to 5%. Nevertheless, the traditional BEM method could not differentiate an upwind cone from its downwind counterpart under the same chord, twist and airfoil distributions. Many studies on load-aligned concepts or comparing upwind/downwind designs have utilized tools based on traditional BEM, which arguably makes their conclusions disputable. A correction to the traditional BEM method must be made before it can be used to assess new cone concepts. Such an improved BEM should consider the influence of blade tip position and cone angle, and adopt the corrected inflow velocity decomposition. To date, debate over up- or down-wind coning is ongoing. The design and optimization of super-large coned rotors still has a long way to go.

Author Contributions: Conceptualization, Z.S. and W.Z.; methodology, Z.S. and W.Z.; software, W.S. and W.Z.; validation, Z.S., W.Z. and W.S.; formal analysis, Z.S., W.Z., W.S. and Q.T.; investigation, Z.S.; resources, W.Z. and W.S.; data curation, Z.S., Q.T., J.C., X.L. and W.S.; writing—original draft preparation, Z.S., W.Z. and W.S.; writing—review and editing, Z.S., W.Z. and W.S.; visualization, Z.S.; supervision, W.Z. and W.S.; project administration, W.Z.; funding acquisition, Z.S. and W.Z. All authors have read and agreed to the published version of the manuscript.

Funding: This research was funded by the National Nature Science Foundation of China under grant number 51905469 and 11672261; the National key research and development program of China under grant number 2019YFE0192600; the Nature Science Foundation of Yangzhou under grant number YZ2019074; an open funding from Shanxi Key Laboratory of Industrial Automation under grant number SLGPT2019KF01-13.

Institutional Review Board Statement: Not applicable.

Informed Consent Statement: Not applicable.

Data Availability Statement: Data available on request due to restrictions eg privacy or ethical. The data presented in this study are available on request from the corresponding author. The data are not publicly available due to the large volume of CFD data.

Conflicts of Interest: The authors declare no conflict of interest.

References

1. Fingersh, L.; Hand, M.; Laxson, A. *Wind Turbine Design Cost and Scaling Model*; Office of Scientific and Technical Information (OSTI): Oak Ridge, TN, USA, 2006.
2. Loth, E.; Steele, A.; Ichter, B.; Selig, M.; Moriarty, P.J. Segmented Ultralight Pre-Aligned Rotor for Extreme-Scale Wind Turbines. In Proceedings of the 50th AIAA Aerospace Sciences Meeting including the New Horizons Forum and Aerospace Exposition, Nashville, TN, USA, 9–12 January 2012. [CrossRef]
3. Barlas, T.K.; Van Kuik, G.A.M. Review of state of the art in smart rotor control research for wind turbines. *Prog. Aerosp. Sci.* **2010**, *46*, 1–27. [CrossRef]
4. Barnes, R.; Morozov, E. Structural optimisation of composite wind turbine blade structures with variations of internal geometry configuration. *Compos. Struct.* **2016**, *152*, 158–167. [CrossRef]
5. Chen, J.; Wang, Q.; Shen, W.; Pang, X.; Li, S.; Guo, X. Structural optimization study of composite wind turbine blade. *Mater. Des.* **2013**, *46*, 247–255. [CrossRef]
6. Sun, Z.; Sessarego, M.; Chen, J.; Shen, W. Design of the OffWindChina 5 MW Wind Turbine Rotor. *Energies* **2017**, *10*, 777. [CrossRef]
7. Crawford, C.; Platts, J. Updating and Optimization of a Coning Rotor Concept. In Proceedings of the 44th AIAA Aerospace Sciences Meeting and Exhibit, Reno, Nevada, 9–12 January 2006; pp. 9–12. [CrossRef]
8. Steele, A.; Ichter, B.; Qin, C.; Loth, E.; Selig, M.; Moriarty, P. Aerodynamics of an Ultra light Load-Aligned Rotor for Extreme-Scale Wind Turbines. In Proceedings of the 51st AIAA Aerospace Sciences Meeting Including the New Horizons Forum and Aerospace Exposition, Grapevine, TX, USA, 7–10 January 2013.
9. Noyes, C.; Qin, C.; Loth, E. Pre-aligned downwind rotor for a 13.2 MW wind turbine. *Renew. Energy* **2018**, *116*, 749–754. [CrossRef]
10. Noyes, C.; Qin, C.; Loth, E. Analytic analysis of load alignment for coning extreme-scale rotors. *Wind. Energy* **2020**, *23*, 357–369. [CrossRef]
11. Qin, C.; Loth, E.; Zalkind, D.S.; Pao, L.Y.; Yao, S.; Griffith, D.T.; Selig, M.S.; Damiani, R. Downwind coning concept rotor for a 25 MW offshore wind turbine. *Renew. Energy* **2020**, *156*, 314–327. [CrossRef]
12. Wanke, G.; Bergami, L.; Larsen, T.J.; Hansen, M.H. Changes in design driving load cases: Operating an upwind turbine with a downwind rotor configuration. *Wind. Energy* **2019**, *22*, 1500–1511. [CrossRef]
13. Bortolotti, P.; Kapila, A.; Bottasso, C.L. Comparison between upwind and downwind designs of a 10 MW wind turbine rotor. *Wind. Energy Sci.* **2019**, *4*, 115–125. [CrossRef]
14. Ning, A.; Petch, D. Integrated design of downwind land-based wind turbines using analytic gradients. *Wind. Energy* **2016**, *19*, 2137–2152. [CrossRef]
15. Mikkelsen, R.; Sørensen, J.N.; Shen, W.Z. Modelling and analysis of the flow field around a coned rotor. *Wind. Energy* **2001**, *4*, 121–135. [CrossRef]
16. Madsen, H.A.; Bak, C.; Døssing, M.; Mikkelsen, R.F.; Øye, S. Validation and modification of the Blade Element Momentum theory based on comparisons with actuator disc simulations. *Wind. Energy* **2010**, *13*, 373–389. [CrossRef]
17. Crawford, C. Re-examining the precepts of the blade element momentum theory for coning rotors. *Wind. Energy* **2006**, *9*, 457–478. [CrossRef]

18. Madsen, H.A.; Rasmussen, F. The influence on energy conversion and induction from large blade deflections. In Proceedings of the 1999 European Wind Energy Conference and Exhibition, Nice, France, 1–5 March 1999.
19. Farhan, A.; Hassanpour, A.; Burns, A.; Motlagh, Y.G. Numerical study of effect of winglet planform and airfoil on a horizontal axis wind turbine performance. *Renew. Energy* **2019**, *131*, 1255–1273. [CrossRef]
20. Chattot, J.-J. Effects of blade tip modifications on wind turbine performance using vortex model. *Comput. Fluids* **2009**, *38*, 1405–1410. [CrossRef]
21. Shen, X.; Chen, J.-G.; Zhu, X.-C.; Liu, P.-Y.; Du, Z.-H. Multi-objective optimization of wind turbine blades using lifting surface method. *Energy* **2015**, *90*, 1111–1121. [CrossRef]
22. Xie, W.; Zeng, P.; Lei, L. A novel folding blade of wind turbine rotor for effective power control. *Energy Convers. Manag.* **2015**, *101*, 52–65. [CrossRef]
23. Kress, C.; Chokani, N.; Abhari, R. Downwind wind turbine yaw stability and performance. *Renew. Energy* **2015**, *83*, 1157–1165. [CrossRef]
24. Kress, C.; Chokani, N.; Abhari, R.S. Design Considerations of rotor cone angle for downwind wind turbines. In Proceedings of the ASME Turbo Expo 2015: Turbine Technical Conference and Exposition, Montreal, QC, Canada, 15–19 June 2015. [CrossRef]
25. Sun, Z.; Zhu, W.; Shen, W.; Zhong, W.; Cao, J.; Tao, Q. Aerodynamic Analysis of Coning Effects on the DTU 10 MW Wind Turbine Rotor. *Energies* **2020**, *13*, 5753. [CrossRef]
26. Zahle, F.; Bak, C.; Sørensen, N.N.; Guntur, S.; Troldborg, N. Comprehensive Aerodynamic Analysis of a 10 MW Wind Turbine Rotor Using 3D CFD. In Proceedings of the 32nd ASME Wind Energy Symposium, National Harbor, MD, USA, 13–17 January 2014. [CrossRef]
27. The DTU 10MW Reference Wind Turbine Project Site. Available online: https://rwt.windenergy.dtu.dk/dtu10mw/dtu-10mw-rwt (accessed on 6 November 2018).
28. Jost, E.; Lutz, T.; Krämer, E. Steady and Unsteady CFD Power Curve Simulations of Generic 10 MW Turbines. In Proceedings of the 11th EAWE PhD Seminar on Wind Energy in Europe, Stuttgart, Germany, 23–25 September 2015.
29. Menter, F.R. Two-equation eddy-viscosity turbulence models for engineering applications. *AIAA J.* **1994**, *32*, 1598–1605. [CrossRef]
30. Sørensen, N.N. General Purpose Flow Solver Applied to Flow over Hills. Ph.D. Thesis, Technical University of Denmark, Lyngby, Denmark, September 1995.
31. Jost, E.; Klein, L.; Leipprand, H.; Lutz, T.; Krämer, E. Extracting the angle of attack on rotor blades from CFD simulations. *Wind. Energy* **2018**, *21*, 807–822. [CrossRef]
32. Rahimi, H.; Schepers, J.; Shen, W.; García, N.R.; Schneider, M.; Micallef, D.; Ferreira, C.S.; Jost, E.; Klein, L.; Herráez, I. Evaluation of different methods for determining the angle of attack on wind turbine blades with CFD results under axial inflow conditions. *Renew. Energy* **2018**, *125*, 866–876. [CrossRef]
33. Hansen, M.O.; Sørensen, N.N.; Michelsen, J. Extraction of lift, drag and angle of attack from computed 3-D viscous flow around a rotating blade. In Proceedings of the 1997 European Wind Energy Conference, Dublin, Ireland, 6–9 October 1997.

Article

The Influence of Tilt Angle on the Aerodynamic Performance of a Wind Turbine

Qiang Wang [1], Kangping Liao [1,*] and Qingwei Ma [2]

[1] College of Shipbuilding Engineering, Harbin Engineering University, Harbin 150001, China; wangqiang918@hrbeu.edu.cn
[2] School of Mathematics, Computer Sciences & Engineering, City, University of London, London EC1V 0HB, UK; q.ma@city.ac.uk
* Correspondence: liaokangping@hrbeu.edu.cn; Tel.: +86-451-8256-8147

Received: 16 May 2020; Accepted: 1 August 2020; Published: 4 August 2020

Featured Application: This research has certain reference significance for improved wind turbine performance. It can also provide reference value for wind shear related study.

Abstract: Aerodynamic performance of a wind turbine at different tilt angles was studied based on the commercial CFD software STAR-CCM+. Tilt angles of 0, 4, 8 and 12° were investigated based on uniform wind speed and wind shear. In CFD simulation, the rotating motion of blade was based on a sliding mesh. The thrust, power, lift and drag of the blade section airfoil at different tilt angles have been widely investigated herein. Meanwhile, the tip vortices and velocity profiles at different tilt angles were physically observed. In addition, the influence of the wind shear exponents and the expected value of turbulence intensity on the aerodynamic performance of the wind turbine is also further discussed. The results indicate that the change in tilt angle changes the angle of attack of the airfoil section of the wind turbine blade, which affects the thrust and power of the wind turbine. The aerodynamic performance of the wind turbine is better when the tilt angle is about 4°. Wind shear will cause the thrust and power of the wind turbine to decrease, and the effect of the wind shear exponents on the aerodynamic performance of the wind turbine is significantly greater than the expected effect of the turbulence intensity. The main purpose of the paper was to study the effect of tilt angle on the aerodynamic performance of a fixed wind turbine.

Keywords: wind turbine; tilt angle; unsteady aerodynamics; computational fluid dynamics

---

1. Introduction

The use of wind energy has increased over the past few decades. Today, wind energy is the fastest growing renewable energy source in the world [1]. Despite the amazing growth in the installed capacity of wind turbines in recent years, engineering and science challenges still exist [2]. The main goals in wind turbine optimization are to improve wind turbine performance and to make them more competitive on the market. Studies have shown that the wind turbine tilt angle affects the shear force and bending moment at the tower top and the blade root [3], and the interaction between the blade and the tower also affects the aerodynamic performance of the wind turbine [4]. Therefore, it is necessary to study the effect of tilt angle on wind turbine performance and analyze the characteristics of blade–tower interaction, aiming to improve the wind turbine performance.

In recent years, more and more scholars have been paying attention to the interaction between the blades and towers of wind turbines. Kim et al. [4] studied the interaction between the blade and the tower using the nonlinear vortex correction method. They concluded that as the yaw angle and wind shear exponent increase, the interaction between the blade and the tower decreases. The influence of the tower diameter on the interaction between the blades and the tower is higher than that of the tower clearance. Meanwhile, this interaction may increase the total fatigue load at low wind speed. Guo et al. [5] used blade element moment (BEM) theory to study the interaction between the blade and tower. Their results show that the blade–tower interaction is much more significant than that of the wind shear. Wang et al. [6] researched the blade–tower interaction using computational fluid dynamics (CFD). Their research shows that the influence of the tower on the total aerodynamic performance of the upwind wind turbine is small, but the rotating blade will cause an obvious periodic drop in the front pressure of the tower. At the same time, we can see the strong interaction of blade tip vortices. Narayana et al. [7] researched the gyroscopic effect of small-scale wind turbines. Their findings show that changing the tilt angle can improve the aerodynamic performance of small-scale wind turbines. Recently, Zhao et al. [3] proposed a new wind turbine control method. In their control method, tilt angle increases as wind speed increases, with the purpose of reducing the blade loading and maintaining the power of the wind turbine at high wind speeds. Their research shows that the new control method can reduce the shear force at the top and bottom of the tower when compared with the yaw control strategy.

Many researchers have studied the effect of tilt angle on the structural performance of a wind turbine. For example, Zhao et al. [8] studied the structural performance of a two-blade downwind wind turbine at different tilt angles. However, there is little research on the effect of tilt angle on the aerodynamic performance of wind turbines. In this paper, aerodynamic performance of a wind turbine at different tilt angles is studied. All simulations are performed in CFD software STAR-CCM+ 12.02. Through a comparison of aerodynamic performance of the wind turbine at different tilt angles, the effects of tilt angle on the thrust, power and wake of the wind turbine are studied.

2. Numerical Modeling

2.1. Physical Model

In this study, the governing equation uses the unsteady Reynolds-averaged Navier–Stokes equation. The $SST\ k-\omega$ turbulence model was used in current simulations. A separated flow model was used to solve the flow equation. SIMPLE solution algorithm was used for pressure correction. Convection terms used the second-order upwind scheme. In the unsteady simulation, the time discretization used the second-order central difference scheme. In addition, due to the sliding mesh approach, no hole cutting was necessary, making the calculations more efficient than with the use of an overset mesh. Thus the sliding mesh technique was used to handle rotating motion of a blade [9].

2.2. Turbulence Model

The $SST\ k-\omega$ turbulence model can consider the complex flow of the adverse pressure gradient near the wall region and the flow in the free shear region. Thus, the $SST\ k-\omega$ turbulence model is suitable for simulating the rotational motion of the blade [10]. In addition, this turbulence model can accurately capture wind turbine wake [11,12].

In the Reynolds-averaged N-S equations, $\tau_{ij} = -\overline{\rho u_i' u_j'}$ refers to the Reynolds stress tensor. Reynolds stress tensor and mean strain rate tensor (S_{ij}) are related by the Boussinesq eddy viscosity assumption:

$$\tau_{ij} = 2v_t S_{ij} - \frac{2}{3}\rho k\delta_{ij} \tag{1}$$

where v_t refers to the eddy viscosity, ρ refers to the density, k refers to the turbulence kinetic energy and δ_{ij} refers to the Kronecker delta function.

To provide closure equations, in the $SST\ k-\omega$ turbulence model, the turbulent kinetic energy (k) and specific dissipation of turbulent kinetic energy (ω) also need governing transport equations, which are given as follows:

$$\frac{D\rho k}{Dt} = \tau_{ij}\frac{\partial u_i}{\partial x_j} - \beta^*\rho\omega k + \frac{\partial}{\partial x_j}\left[(\mu + \sigma_k\mu_t)\frac{\partial k}{\partial x_j}\right] \tag{2}$$

$$\frac{D\rho\omega}{Dt} = \frac{\gamma}{\nu_t}\tau_{ij}\frac{\partial u_i}{\partial x_j} - \beta\rho\omega^2 + \frac{\partial}{\partial x_j}\left[(\mu + \sigma_\omega\mu_t)\frac{\partial\omega}{\partial x_j}\right] + 2(1-F_1)\rho\sigma_{\omega2}\frac{1}{\omega}\frac{\partial k}{\partial x_i}\frac{\partial\omega}{\partial x_j} \tag{3}$$

In the formulas above, the model coefficients are defined as follows:

$$\beta^* = F_1\beta_1^* + (1-F_1)\beta_2^* \tag{4}$$

$$\beta = F_1\beta_1 + (1-F_1)\beta_2 \tag{5}$$

$$\gamma = F_1\gamma_1 + (1-F_1)\gamma_2 \tag{6}$$

$$\sigma_k = F_1\sigma_{k1} + (1-F_1)\sigma_{k2} \tag{7}$$

$$\sigma_\omega = F_1\sigma_{\omega1} + (1-F_1)\sigma_{\omega2} \tag{8}$$

The blending function F_1 is defined as follows:

$$F_1 = \tanh\left\{\left\{\min\left[\max\left(\frac{\sqrt{k}}{\beta^*\omega y}, \frac{500\nu_\infty}{y^2\omega}\right), \frac{4\rho\sigma_{\omega2}k}{CD_{k\omega}y^2}\right]\right\}^4\right\} \tag{9}$$

where $CD_{k\omega}$ refers to the cross-diffusion term, y refers to the distance to the nearest wall and ν refers to the kinematic viscosity. F_1 is equal to zero in the region away from the wall ($k-\varepsilon$ turbulence model) and one in the region near the wall ($k-\omega$ turbulence model).

The eddy viscosity is

$$\nu_t = \frac{a_1 k}{\max(a_1\omega, \Omega F_2)} \tag{10}$$

where Ω is the absolute value of the vorticity and F_2 is the second blending function, defined as

$$F_2 = \tanh\left[\max\left(\frac{2\sqrt{k}}{\beta^*\omega y}, \frac{500\nu}{y^2\omega}\right)^2\right] \tag{11}$$

A more detailed description of the *SST* $k-\omega$ turbulence model is provided in [10]. In this study, the parameters for the *SST* $k-\omega$ turbulence model are as follows:

$\sigma_{k1} = 0.85 \quad \sigma_{\omega1} = 0.5 \quad \beta_1 = 0.075 \quad a_1 = 0.31 \quad \beta^* = 0.09 \quad k = 0.41 \quad \sigma_{k2} = 1$

$\sigma_{\omega2} = 0.856 \quad \beta_2 = 0.0828 \quad \gamma_1 = \frac{\beta_1}{\beta^*} - \frac{\sigma_{\omega1}k^2}{\sqrt{\beta^*}} \quad \gamma_2 = \frac{\beta_2}{\beta^*} - \frac{\sigma_{\omega2}k^2}{\sqrt{\beta^*}}$

2.3. Computational Domain

The computational domain was divided into the rotating and outer domains, as shown in Figure 1. The size of the entire outer domain was 12D(x) × 5D(y) × 4D(z). The distance from the wind turbine to the velocity inlet was 3D, and the distance to the pressure outlet was 9D, where D is the diameter of the wind turbine. Due to the complex geometry of the blades, we used the trimmed cell mesh technology to generate high-quality meshes. In order to capture the complex flow around the blade, a fine mesh was used around the blade. A 10-layer boundary layer mesh was generated near the blade and the hub. The total thickness of the boundary layer was 0.03 m, and the growth rate was 1.2. A six-layer boundary layer mesh was generated near the tower and the nacelle. The total thickness of the boundary layer was 0.1 m, and the growth rate was 1.2. Figure 2b shows the refined sliding mesh regions around the blade. Figure 2c,d shows a close-up view of the blades and nacelle tower.

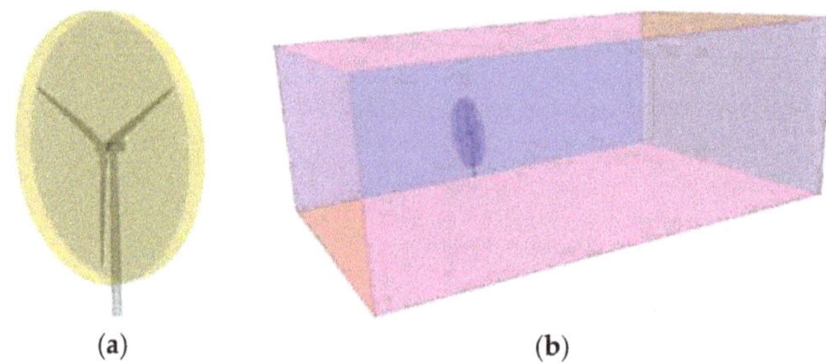

(a) **(b)**

Figure 1. Rotating and outer domain: (**a**) rotation domain for wind turbine simulation; (**b**) entire computational domain for numerical simulation.

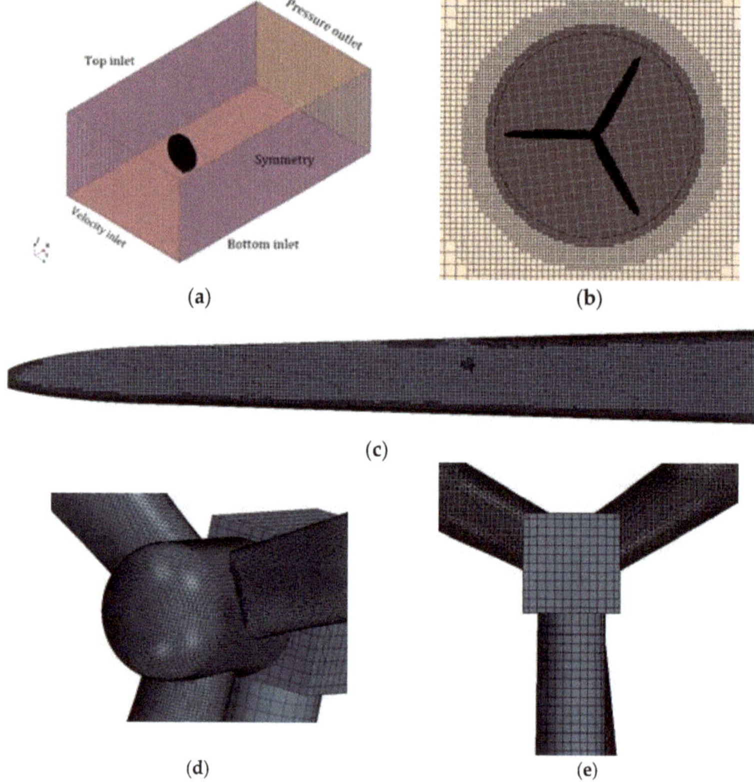

(a) **(b)**

(c)

(d) **(e)**

Figure 2. The computational mesh domain for the wind turbine: (**a**) full grid domain, (**b**) sliding mesh regions, (**c**) close-up view of the blade surface, (**d**) close-up view of the hub surface mesh and (**e**) close-up view of nacelle and tower.

2.4. Boundary Conditions

Figure 2a illustrates the setting of the boundary conditions in this study. In the computational domain, the inlet boundary, bottom and top surfaces were set as velocity inlets. The pressure outlet was set at the outlet boundary. The sides of the computational domain were set to the plane of symmetry. In this simulation, all of the y+ wall treatment of near-wall modeling was applied. In order to reduce the convergence order and improve the solution accuracy, the maximum internal iterations within each time-step was 10 [13].

3. Results and Discussion

3.1. Validations

The 1/75 scale model of a DTU 10 MW reference wind turbine was used for the mesh independence test. In the numerical verification, the tilt angle of the wind turbine was not considered. The main parameters of the scale model are given in Table 1. A detailed introduction of the blade parameters at 40 different blade sections is provided by [14]. Figure 3 shows the wind turbine geometric model and the surface grid. After scaling according to the scale factor, the boundary layers near the blade and hub surface have five layers of refined grid with the total layer thickness of 0.004 m and a progression factor of 1.2.

Table 1. Principal dimensions of the scale model.

Specifications	DTU Down-Scaled
Number of Blades	3
Rotor Diameter (m)	2.37
Hub Diameter (m)	0.178
Rated Wind Speed (m/s)	5.53
Rated Rotor Speed (rpm)	330

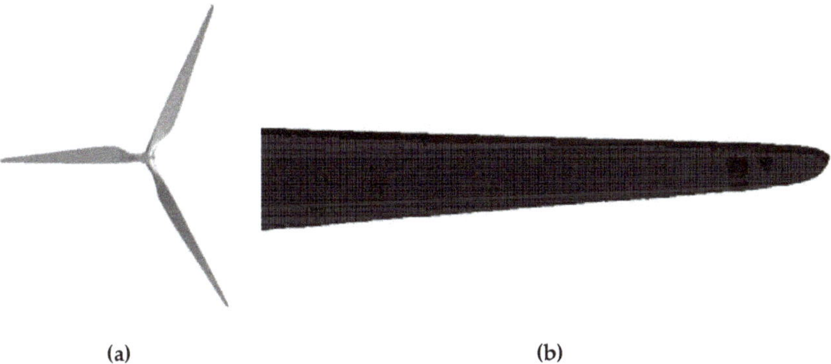

(a)   (b)

Figure 3. Geometric model and surface grid: (**a**) the rotor geometric model; (**b**) the blade surface grid.

The blade surface mesh size includes the maximum mesh size and the minimum mesh size. The number of meshes corresponding to different mesh sizes is shown in Table 2. According to previous study, the time-step size corresponding to 1° increment of azimuth angle of the wind turbine per time-step was applied in all simulations [15]. Moreover, the simulation was run under unsteady conditions. The comparison of thrust and torque for different grid resolutions with the same wind speed of 5.53 m/s, rotor speed of 330 rpm and time-step size of 5×10^{-4} s is presented in Tables 3 and 4. It can be observed from Tables 3 and 4 that the grid resolution of Case 3 is sufficient to solve the unsteady aerodynamics of the wind turbine. Therefore, the grid resolution of Case 3 was used in subsequent simulations.

Table 2. Mesh size of blade surface.

CFD Mesh Type	Case 1	Case 2	Case 3	Case 4
Maximum Size (mm)	3.000	2.000	1.500	1.100
Minimum Size (mm)	0.500	0.350	0.250	0.180
Total Mesh Number (million)	1.850	3.240	4.630	9.400

Table 3. Comparison of thrust between experiment and CFD simulation at different grid densities.

CFD Mesh Type	LIFES50+ Wind Tunnel Data (N), [14]	Present Study (N)	Error (%)
Case 1		70.010	2.000
Case 2	68.631	69.660	1.500
Case 3		69.520	1.300
Case 4		69.500	1.300

Table 4. Comparison of torque between experiment and CFD simulation at different grid densities.

CFD Mesh Type	LIFES50+Wind Tunnel Data ($N{\cdot}M$), [14]	Present Study ($N{\cdot}M$)	Error (%)
Case 1		5.690	8.700
Case 2	6.232	5.850	6.100
Case 3		5.900	5.300
Case 4		5.920	5.000

Simulations at different wind speeds were performed, and the simulation results were compared with wind tunnel experiment data, as presented in Figure 4. In this paper, we always keep the pitch angle at 0°, so we have not considered the working conditions above the rated wind speed. When the wind speed is close to the rated wind speed, the thrust and torque of the CFD simulation are lower than those of the wind tunnel experiment, but the maximum error is not more than 10%. This means that STAR-CCM+ can accurately simulate the aerodynamic performance of the wind turbine under rotating motion.

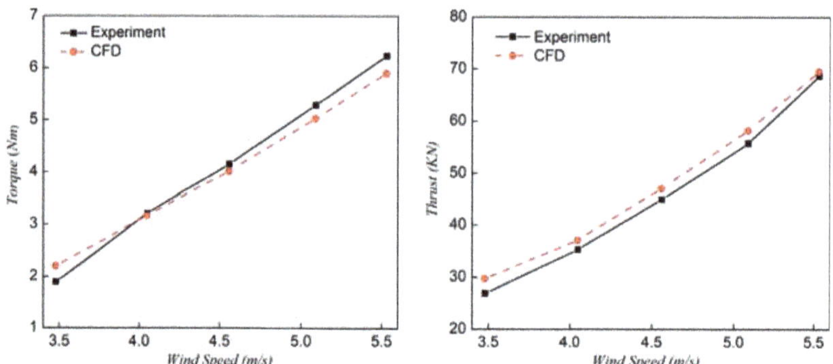

Figure 4. Comparison of thrust and torque between wind tunnel experiment and CFD simulation at different wind speeds (Case 3).

In order to ensure the reliability of the NREL 5 MW real-scale wind turbine simulation, the NREL 5 MW real-scale wind turbine was used for grid convergence analysis. Major properties of the NREL 5 MW reference wind turbine are given in Table 5 [16]. Figure 5 shows the blade alone geometric model and full configuration geometric model with the tower. The blade alone model was used for numerical verification, and the full configuration model was used to investigate the effect of tilt angle on the aerodynamic performance of the wind turbine. Near the wall surface of the blades and hub, the boundary layers have 10 layers of refined grid with the total layer thickness of 0.03 m and a progression factor of 1.2. The same wind speed of 11.4 m/s and rotor speed of 12.1 rpm were applied in all simulations. Meanwhile, in all simulations, the time step is the time taken by the wind turbine to increase the azimuth angle by 1°.

Table 5. Principal dimensions of the NREL 5 MW reference wind turbine.

Specifications	
Rated Power (MW)	5
Rotor Orientation, Configuration	Upwind, 3 blades
Rated Wind Speed (m/s)	11.4
Rated Rotor Speed (rpm)	12.1
Rotor Diameter (m)	126
Hub Diameter (m)	3
Hub Height (m)	90
Tower Base Diameter (m)	6
Tower Top Diameter (m)	3.87
Pre-cone (°)	2.5

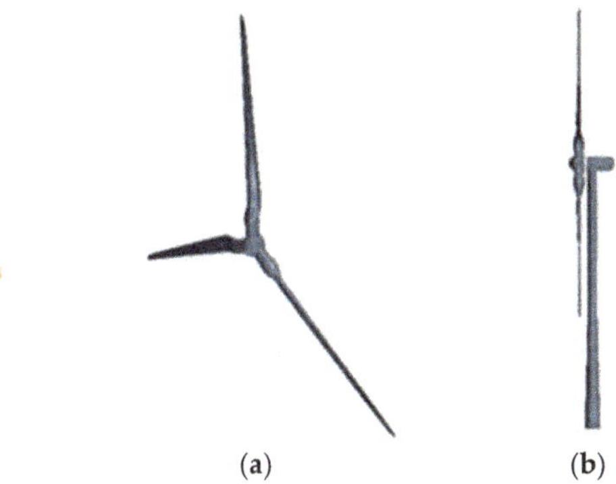

Figure 5. Geometric model of a 5 MW reference wind turbine: (**a**) the rotor geometric model; (**b**) the full configuration model.

The number of meshes corresponding to different mesh sizes is shown in Table 6. The comparison of power for different grid resolutions with the same wind speed of 11.4 m/s and rotor speed of 12.1 rpm is presented in Table 7. It can be observed from Table 7 that the grid resolution of Case 2 is sufficient to solve the unsteady aerodynamics of the wind turbine. Therefore, the grid of Case 2 was used for the simulation of NREL 5 MW real-scale wind turbines at different wind speeds.

Table 6. Mesh size of blade surface.

CFD Mesh Type	Case 1	Case 2	Case 3
Maximum Size (m)	0.20	0.10	0.05
Minimum Size (m)	0.04	0.02	0.01
Total Mesh Number (Million)	1.52	4.80	9.53

Table 7. Comparison of power between NREL data and CFD simulation at different grid densities.

CFD Mesh Type	NREL Data (MW), [16]	Present Study (N)	Error (%)
Case 1		4.767	4.700
Case 2	5.000	4.981	0.380
Case 3		5.020	0.400

Aerodynamic simulations of a wind turbine with various wind speeds were tested and compared with the FAST results. The obtained thrust and power were compared with the corresponding NREL data calculated by FAST V8, as presented in Figure 6. The power agrees well with the NREL data, but the thrust tends to be smaller than that from NREL data. The reason for the difference between the CFD method and the FAST can be summarized as follows: (a) the FAST does not consider the three-dimensional flow effects around blades; (b) in the BEM method, in order to calculate a rotor with a limited number of blades, a tip loss correction model needs to be added. The results obtained by different tip loss correction models are also quite different [17]. FAST uses a Prandtl tip loss correction model [16]. Therefore, the CFD result of the thrust is significantly lower than the FAST result. A similar phenomenon appeared in [18]. However, at the rated wind speed, compared with FAST data, the errors of the thrust and power obtained by CFD are less than 5% Through the above analysis, the grid of Case 2 can accurately simulate the aerodynamic performance of NREL 5 MW real-scale wind turbines. Therefore, the grid of Case 2 was used to simulate the effect of tilt angle on the aerodynamic performance of the wind turbine.

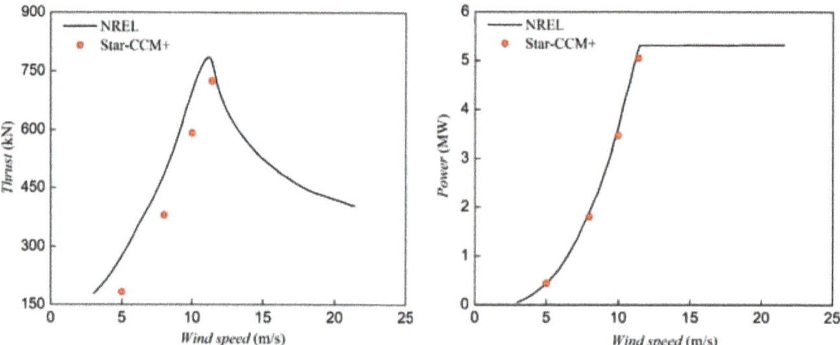

Figure 6. Comparisons of thrust and power.

3.2. The Effect of the Tilt Angle on the Aerodynamic Performance of the Wind Turbine

In this study, nacelle tilt angles of 0, 4, 8 and 12° were investigated. Figure 7 shows the structure of the wind turbine at different tilt angles. In the picture, β is the pre-coning angle, and γ is the shaft tilt angle. The azimuth of the rotor is defined as ψ, as presented in Figure 8. In Figure 8, the blue rotor is the initial position with the 0-azimuth angle. Subsequent analysis is based on the results after the wind turbine has stabilized. Under different tilt angles, the change in wind turbine thrust and power with the azimuth is shown in Figure 9. Comparing the no-tower curve with the other four curves, it can be seen that the thrust and power generate periodic fluctuations due to the influence of the tower. When the blades pass through the tower, the thrust and power will periodically decrease. This phenomenon is called the blade–tower interaction (BTI) [19]. The BTI effects begin at approximately 30° rotor azimuth and dissipate at approximately 100° rotor azimuth, as presented in Figure 10. This agrees with previous studies, which all show effects in approximately this same 70° range [19].

Figure 9 shows the difference between the thrust and power at approximately 60, 180 and 300° azimuth with the same nacelle tilt. This phenomenon is due to the interaction between the blade and the tower creating a random vortex. As the nacelle tilt increases, the blade and tower interactions gradually weaken. Therefore, this phenomenon becomes less important as the nacelle tilt increases. In Figure 10, when ψ is approximately 65°, the thrust and power of the wind turbine at 4 and 8° nacelle tilt are higher than 0 and 12° nacelle tilt.

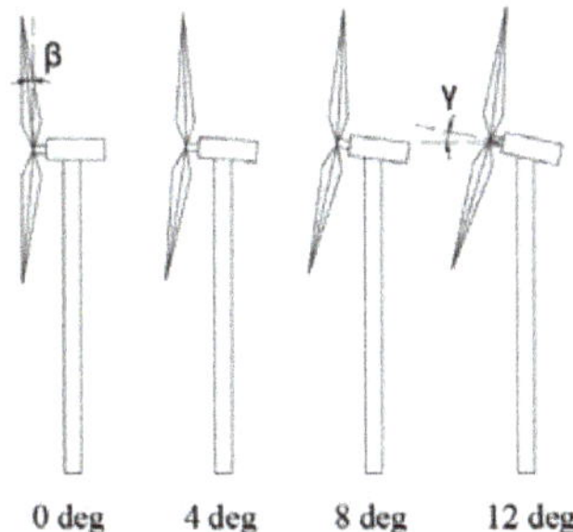

Figure 7. Structure of the wind turbine at different tilt angles.

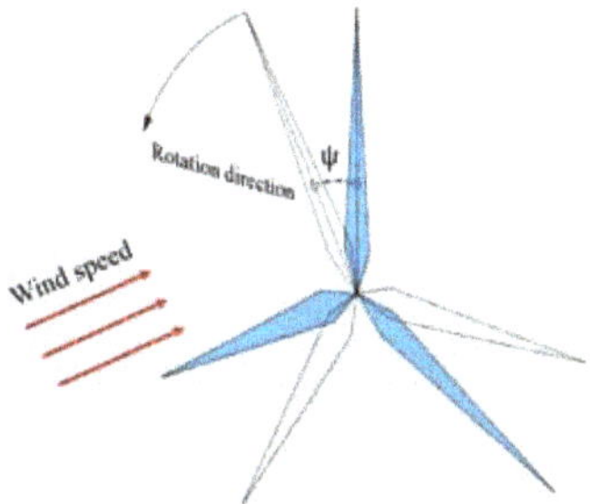

Figure 8. Definition of the azimuth.

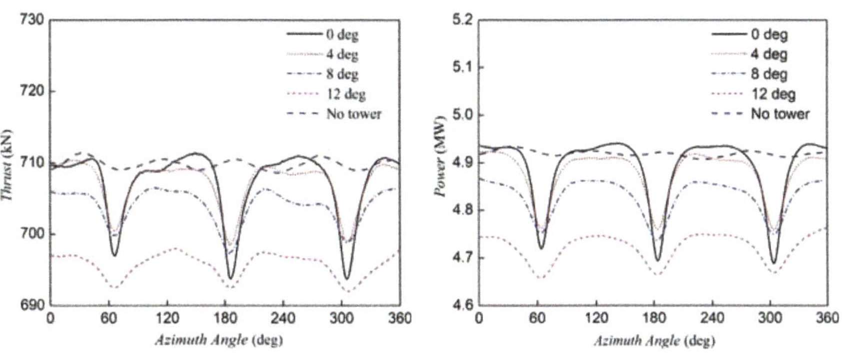

Figure 9. Comparison of thrust and power at different nacelle tilt angles.

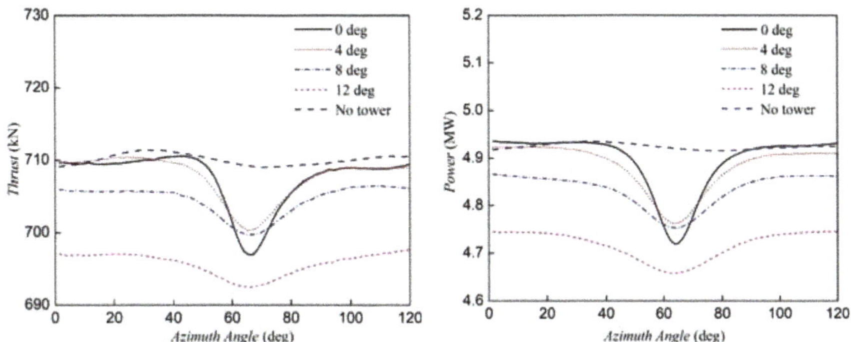

Figure 10. Comparison of thrust and power at different nacelle tilt angles (partial enlargement).

The position of the blade relative to the tower with the 60° azimuth is shown in Figure 11. Instantaneous pressure magnitude and streamlines at blade sections r/R = 0.5, r/R = 0.7 and r/R = 0.9 (Blade 1) of the wind turbine are presented in Figures 12–14. In the low span (r/R = 0.5) suction side, the flow separation phenomenon can be observed. However, the flow remains attached for higher radial sections (r/R = 0.7 and r/R = 0.9). In addition, with the increase of the nacelle tilt, the flow separation of the low span suction side is gradually weakened. The variation of the pressure distribution around different sections airfoil with the nacelle tilt can also be observed.

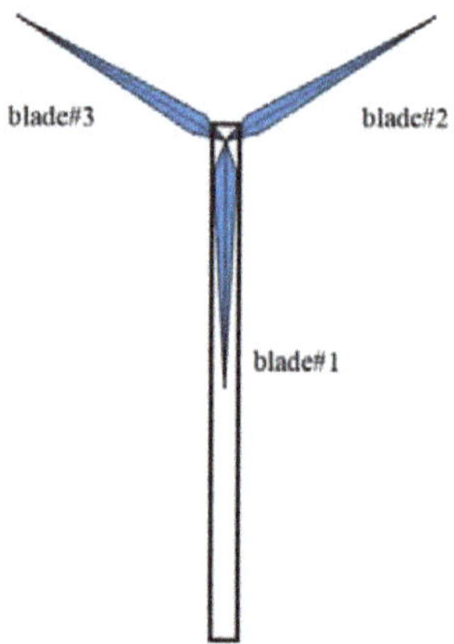

Figure 11. Blade position ($\psi = 60°$).

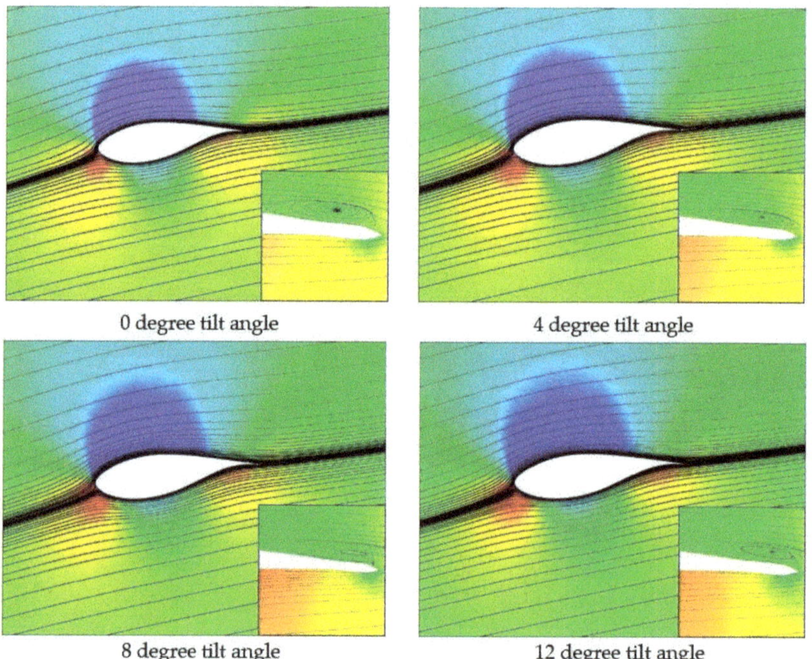

Figure 12. Instantaneous pressure magnitude and streamlines diagram (r/R = 0.5).

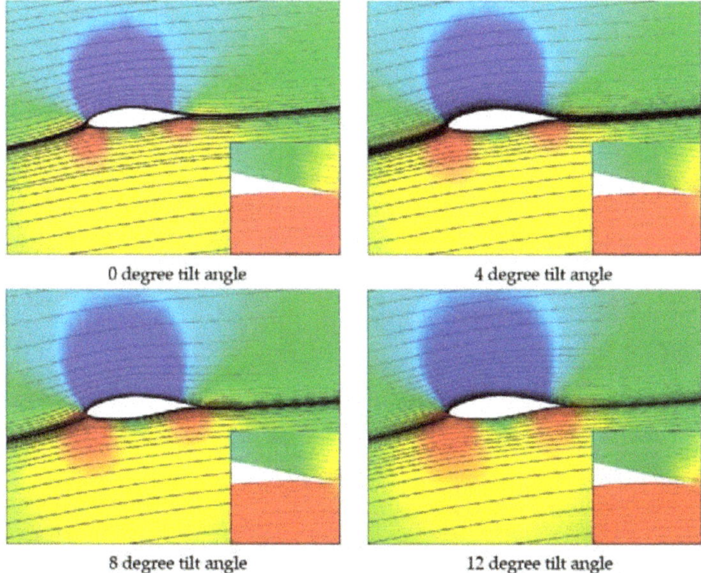

Figure 13. Instantaneous pressure magnitude and streamlines diagram (r/R = 0.7).

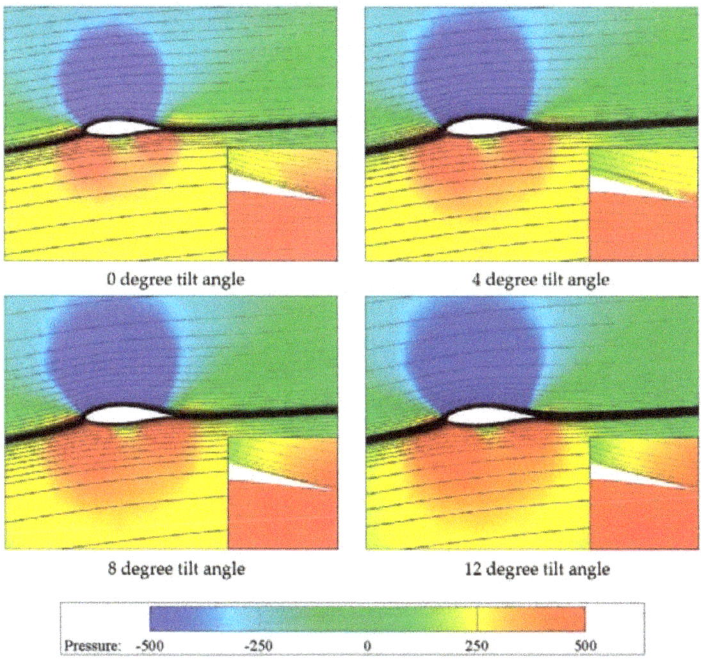

Figure 14. Instantaneous pressure magnitude and streamlines diagram (r/R = 0.9).

H. Rahimi et al. [20] studied different methods of calculating the angle of attack of the wind turbine section airfoil. However, in CFD, when considering the interaction between the blade and the tower, it is difficult to calculate the angle of attack of the blade section airfoil. Therefore, only the effect of tilt angle on the blade section airfoil load is considered in this paper. Figure 15 shows the distribution of azimuth average thrust and tangential force along the blade span. From the figure, we can see that in terms of thrust, when the tilt angle is 4°, the distribution of the thrust along the blade span does not change much compared to the 0° tilt angle. However, when the tilt angle is increased to 8 and 12°, the thrust of the section airfoil at the blade tip is lower than the values at 0 and 4° tilt angle. In terms of tangential force, the tangential force gradually decreases as the tilt angle increases, for up to 0.5 of the span. However, the tangential force at 4° tilt angle does not change much compared to 0° tilt angle. Figure 16 shows the distribution of thrust and tangential force along the blade span when the blade is located in front of the tower. In terms of thrust, the thrust of the section airfoil gradually increases as the tilt angle increases, for up to 0.7 of the span. Regarding the tangential force, the increase of the tilt angle also increases the tangential force, for up to 0.6 of the span. However, regardless of thrust or tangential force, the value at 8° of tilt does not change much compared to 12° of tilt. This means that the influence of the tower becomes weaker after the tilt angle exceeds 8°.

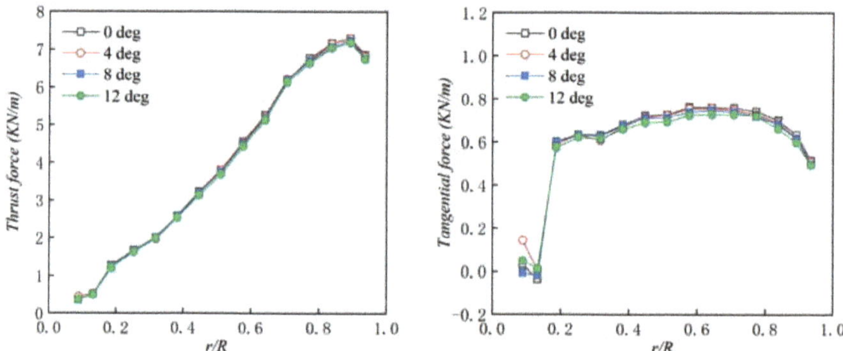

Figure 15. The average thrust and average tangential force per unit of span along the blade span for Blade 1.

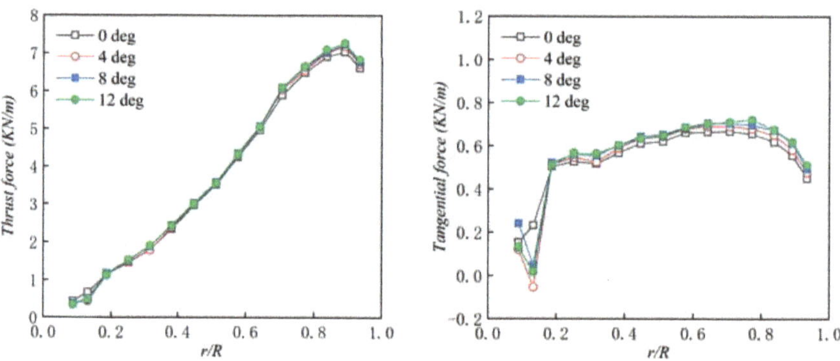

Figure 16. Thrust and tangential force per unit of span along the blade span for Blade 1.

Thrust force per unit of span along the rotor span for Blade 1 is shown in Figure 17. In the blade root, the thrust will fluctuate with the change of the azimuth angle, which is mainly caused by the three-dimensional flow of the blade root. In the middle of the blade, when the azimuth angle is 0-180°, the thrust is the largest at 4° tilt angle, and the thrust is the smallest at 12° tilt angle. When the azimuth angle is 180-360°, the thrust gradually decreases as the tilt angle increases. In the vicinity of the blade tip, when the azimuth angle is 0-180°, except for the tilt angle of 0°, the thrust has a change that increases first and then decreases with the change of the azimuth angle. When the azimuth angle is 180-360°, the thrust curve decreases first and then increases, and the thrust gradually decreases as the elevation angle increases. Figure 18 shows the tangential force per unit of span along the rotor span for Blade 1. We can see that in the middle of the blade and near the tip of the blade, the tangential force of the section airfoil in the 180-360° angle range is higher than the value in the 0–180° angle range, except for the case of the 0° tilt angle. At the same time, we found that in the middle of the blade and near the tip of the blade, when the azimuth angle is 180°, the thrust and tangential force at 0° tilt are the smallest, which is mainly due to the maximum interaction between the blade and the tower at 0° tilt angle.

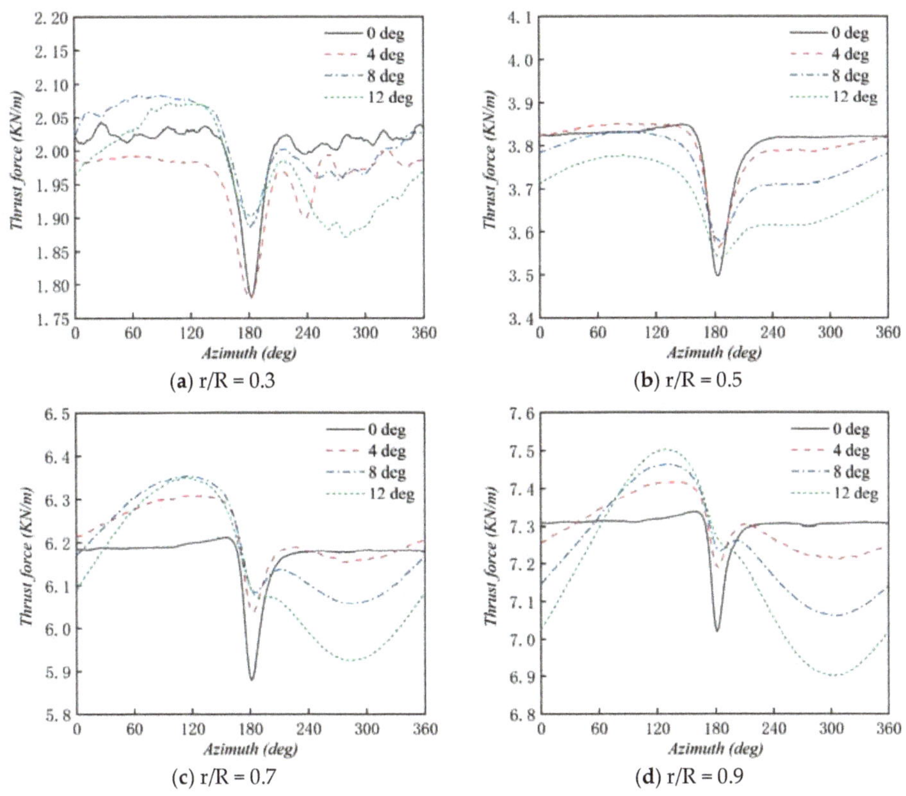

Figure 17. Thrust force per unit of span along the rotor span for Blade 1.

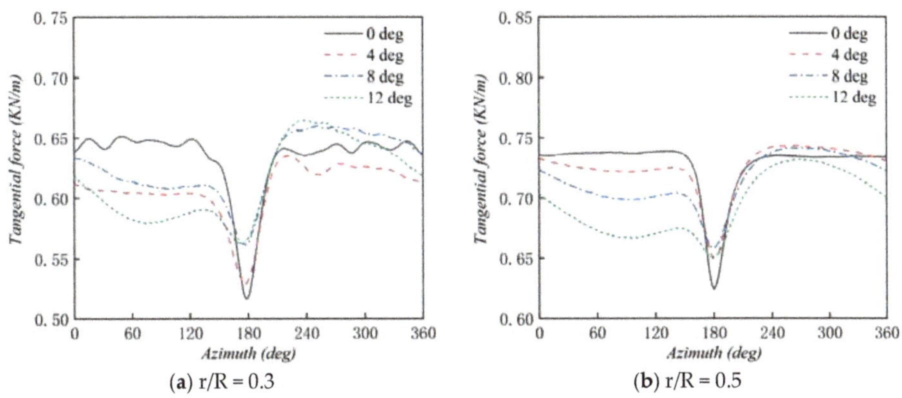

Figure 18. *Cont.*

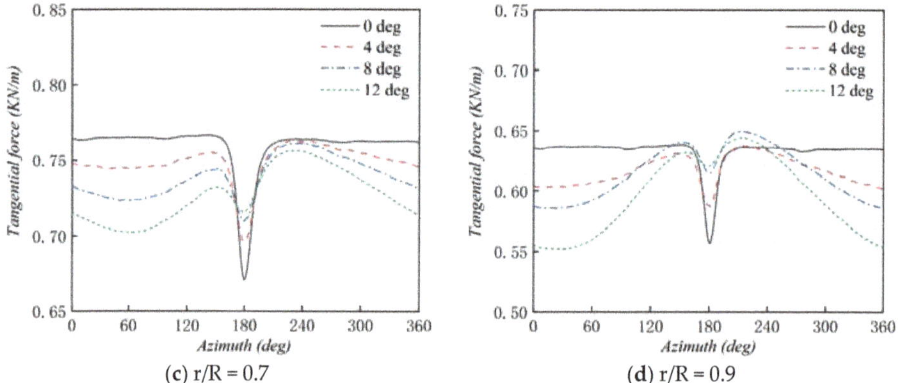

(c) r/R = 0.7   (d) r/R = 0.9

Figure 18. Tangential force per unit of span along the rotor span for Blade 1.

3.3. The Effect of the Tilt Angle on the Wind Turbine Wake

The instantaneous isovorticities occurring when the blade is in front of the tower are presented in Figure 19. One can clearly see that these instantaneous diagrams with nacelle tilt angle shows that there is a strong flow interaction between the wake generated by the blade root, hub and tower regions. Because of the existence of the tower, there are strong unsteady flow interactions between tower vortex and blade tip vortex during downstream propagation. This interaction caused the blade tip vortex to break behind the tower. In addition, an increase in tilt angle will cause the blade tip vortex tube to tilt.

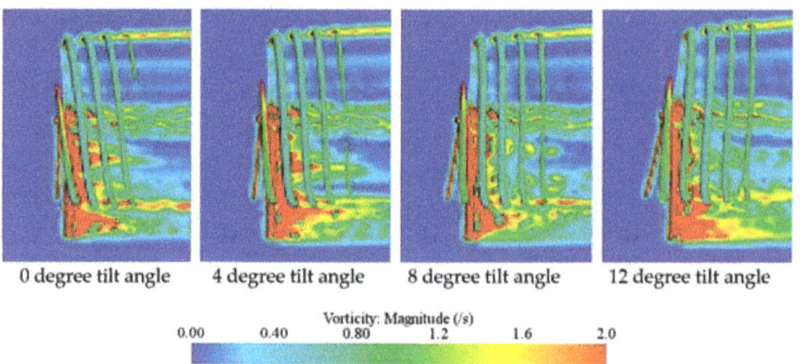

Figure 19. Side-view of instantaneous isovorticity contours for different nacelle tilt angles.

The instantaneous x-vorticities at different sections in four tilt angles are presented in Figure 20. We can observe that there is a clear difference in the blade tip vortex at different tilt angles. At 0 and 4° tilt angles, the blade tip vortex has only negative x-vorticities. When the tilt angle is changed to 8°, positive x-vorticity and negative x-vorticity appear in the right half of the blade tip vortex. When the tilt angle is changed to 12°, the left part of blade tip vortex is negative and right part is positive. At the same time, it can be seen that there are slight differences in the tower-generated vortexes of the four cases. By comparison, at the positions of x/D = 0.25 and x/D = 0.5, the vortex generated by the tower behind the rotor at the tilt angle of 4° is slightly less than other cases. When the tilt angle reaches 12°, the vortex generated by the tower is broken.

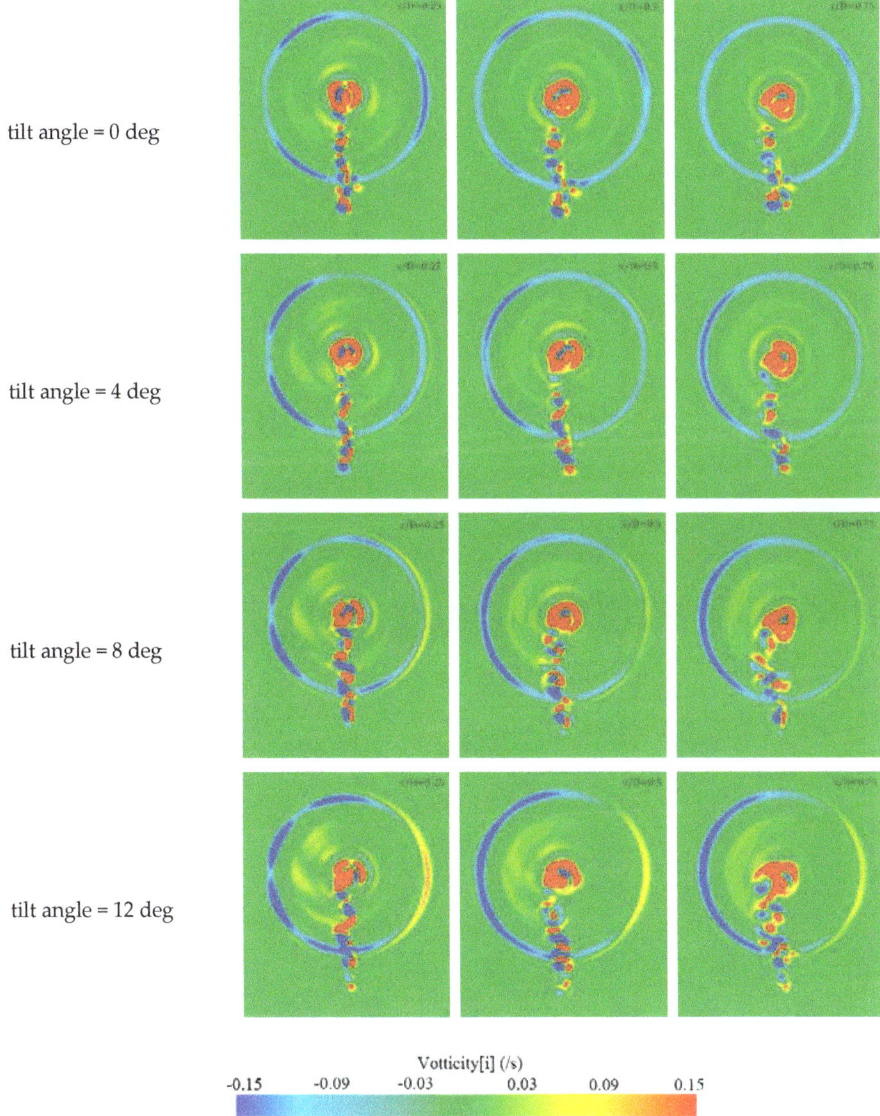

Figure 20. Instantaneous x-vorticities at different sections for four tilt angles.

The corresponding vertical x-velocity profiles are presented in Figure 21. When the tilt angle is 0°, as the downstream distance increases, the velocity field behind the wind turbine is approximately symmetrical about the centerline and keeps a circular shape. However, as the tilt angle increases, the velocity field behind the wind turbine shows asymmetry and gradually moves to the upper right. Meanwhile, the low-velocity region at the end of the wake gradually decreases with increasing tilt angle.

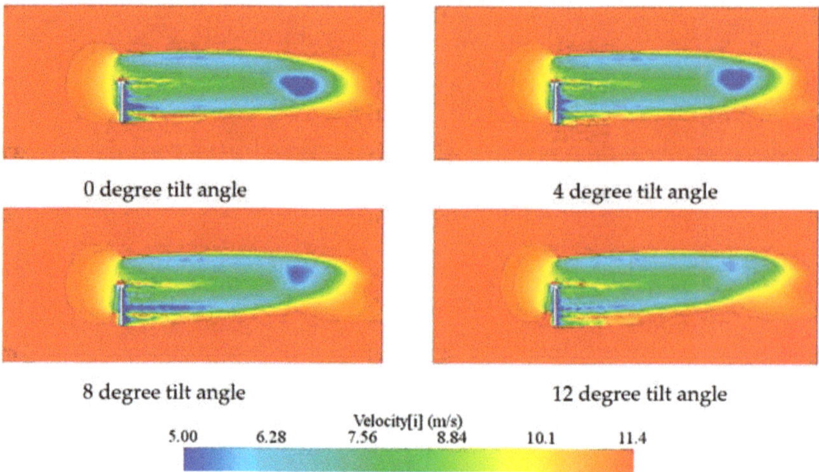

Figure 21. Vertical section x-velocity profiles at y = 0 m.

Figure 22 shows the distribution of instantaneous axial velocity along blade span at the wind turbine downstream positions of 0.5D, 2.5D, 3.5D and 4.5D, which represent the development of the velocity in the wake. Observing the instantaneous axial velocity distribution at the position of X/D = 0.5, it can be seen that the upper half of the curve does not change much with the tilt angle, but the lower half of the curve changes significantly with the tilt angle. In addition, it can be seen that the lower half of the curve has the smallest fluctuation at the 4° tilt angle, which means that the interaction between the blade tip vortex downstream of the wind turbine and the tower wake vortex is the weakest at a tilt angle of 4°. We can also observe a similar phenomenon in Figure 21. Observing the instantaneous axial velocity distribution at the positions of X/D = 2.5 and X/D = 3.5, we can see that as the tilt angle increases, the minimum velocity in the wake gradually increases and shifts upwards. However, at the position of X/D = 4.5, there is a slight decrease in the minimum velocity as the tilt angle increases. This is due to the upward shift of the wake-end deceleration zone.

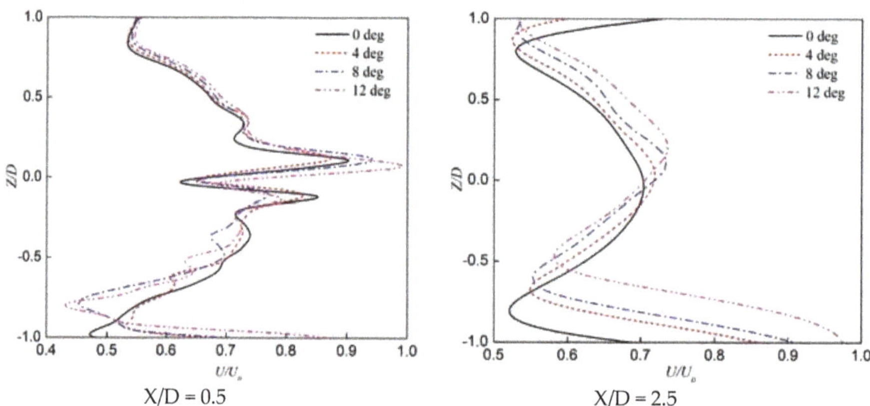

Figure 22. *Cont.*

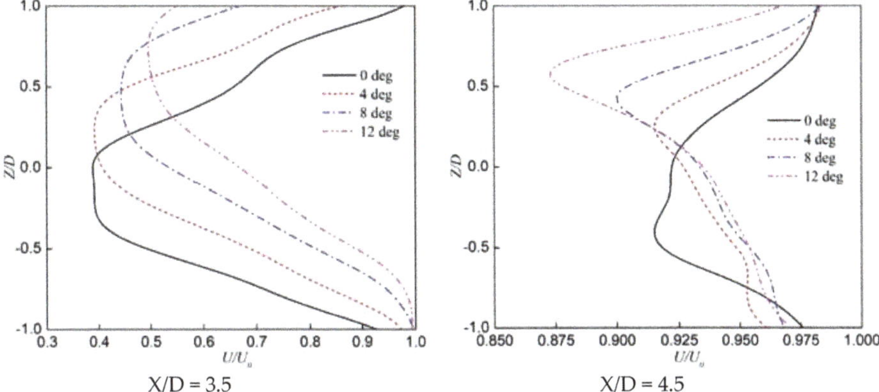

Figure 22. The distribution of instantaneous axial velocity along blade span at the wind turbine downstream positions of 0.5D, 2.5D, 3.5D and 4.5D.

3.4. Wind Shear

The change in wind speed with height was determined according to the power function given in International Electrotechnical Commission (IEC) 61400-1 [21] and presented as follows:

$$\frac{V_Z}{V_{Z_r}} = \left(\frac{Z}{Z_r}\right)^{\gamma} \tag{12}$$

where V_Z refers to the wind speed at height z, V_{Z_r} refers to the reference wind speed at height Z_r and γ refers to the wind shear exponent. Z_r refers to the hub height. In this study, the reference wind speed is 11.4 m/s. In this paper, wind shear exponents are 0.09, 0.2 and 0.41. The wind shear exponent of 0.09 indicates a very unstable atmospheric state, 0.20 represents a neutral state and 0.41 represents a very stable state [4].

The turbulence intensity was calculated according to the formula in IEC 61400-1 [21] and given as follows:

$$I_T = I_{ref}(0.75V_{hub} + 5.6)/V_{hub} \tag{13}$$

where I_T is the turbulence intensity, I_{ref} is the expected value of the turbulence intensity and V_{hub} is the reference velocity at the hub. In this paper, I_{ref} values are 0.12, 0.14 and 0.16. I_{ref} of 0.12 represents lower turbulence characteristics, 0.14 describes medium turbulence characteristics and 0.16 describes higher turbulence characteristics.

Table 8 shows the average power along one rotation of the wind turbine after it has stabilized. It can be seen from Table 8 that, compared with uniform wind, wind shear will cause the average power of the wind turbine to decrease by about 14%. At the same time, it can be found that the average power of the 4° tilt angle is close to that of the 0° tilt angle and is higher than the average power of the 8 and 12° tilt angles under uniform wind or wind shear conditions. The deviation ($|P_a - P_m|$) of the power relative to the average power at an azimuth angle of 180° gradually decreases as the tilt angle increases (see Figure 23, Table 8). When the tilt angle reaches 8° and continues to increase, $|P_a - P_m|$ will remain unchanged. This means that as the tilt angle increases, the interaction between the blade and the tower gradually weakens. When the tilt angle exceeds 4°, the influence of the tilt angle on the interaction between the blade and the tower can be ignored. However, when the tilt angle exceeds 4°, it will cause a significant decrease in the average power of the wind turbine. Therefore, considering the power of the wind turbine and the interaction between the blade and the tower, it is more appropriate to set the wind turbine tilt angle to about 4°.

Table 8. Power for uniform wind and wind shear flow conditions at $V_{hub} = 11.4$ m/s ($\gamma = 0.2$, $I_{ref} = 0.14$).

Tilt Angle (°)	Average Power Pa (MW)			Power Pm at 180° Azimuth Angle (MW)			$|P_a - P_m|$ (MW)	
	Uniform Wind	Wind Shear	Error (%)	Uniform Wind	Wind Shear	Error (%)	Uniform Wind	Wind Shear
0	4.92	4.20	14.63	4.73	4.08	13.74	0.19	0.12
4	4.91	4.21	14.26	4.79	4.15	13.36	0.12	0.06
8	4.85	4.18	13.81	4.78	4.14	13.39	0.07	0.04
12	4.75	4.12	13.26	4.68	4.08	12.82	0.07	0.04

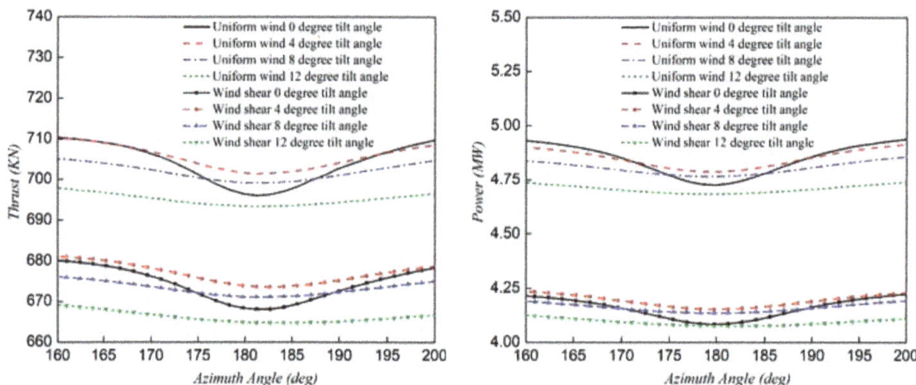

Figure 23. Thrust and power versus azimuth angle for various tilt angles at $V_{hub} = 11.4$ m/s ($\gamma = 0.2$, $I_{ref} = 0.14$).

Figure 24 describes the influence of wind shear exponents (γ) on the aerodynamic performance of the wind turbine. It can be seen from Figure 24 that the thrust and power of the wind turbine when the wind shear exponent is 0.41 are higher than the values when the wind shear exponents are 0.09 and 0.20. It can be found from Table 9 that the average thrust and power of the wind turbine under different wind shear exponents have the smallest error when the wind shear factor is 0.41 compared with the uniform wind, and the average thrust and power of the wind turbine are almost the same when the wind shear factors are 0.09 and 0.2. In Table 9, the wind shear exponent of 0.00 means uniform wind inlet conditions. In Figure 24, it can be seen that the fluctuation of the wind turbine thrust and power curve when the wind shear factor is 0.09 is significantly higher than the other two cases. This means that the wind shear exponent has an effect on the interaction between the blade and the tower.

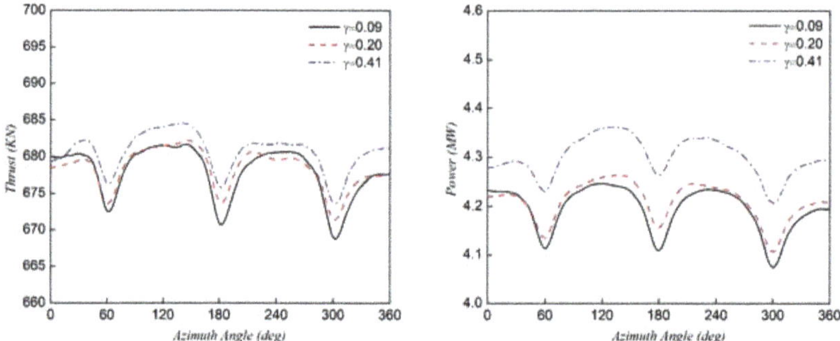

Figure 24. Thrust and power versus azimuth angle for various wind shear exponents (γ) at $V_{hub} = 11.4$ m/s ($I_{ref} = 0.14$, tilt angle = 4°).

Table 9. The power for various wind shear exponents (γ) at V_{hub} = 11.4 m/s (I_{ref} = 0.14, tilt angle = 4°).

Wind Shear Exponents	Average Power Pa (MW)		Average Thrust Ta (KN)	
	Power	Relatively Uniform Wind Error (%)	Thrust	Relatively Uniform Wind Error (%)
0.00	4.91	0.00	709.16	0.00
0.09	4.19	14.66	677.89	4.41
0.20	4.21	14.26	678.38	4.34
0.41	4.30	12.42	680.64	4.02

At the same time, as can be seen from Figure 25, at different turbulence intensity expectations, the thrust and power of the wind turbine are basically the same. This shows that the expected value of the turbulence intensity has little effect on the thrust and power of the wind turbine. Therefore, when using wind shear to simulate a wind turbine, it is necessary to focus on the size of the wind shear exponents according to simulated working conditions.

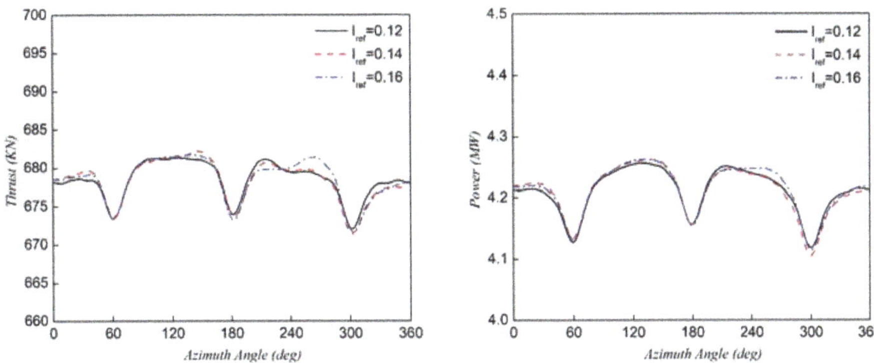

Figure 25. Thrust and power versus azimuth angle for various expected values of the turbulence intensity (I_{ref}) at V_{hub} = 11.4 m/s (γ = 0.2, tilt angle = 4°).

4. Conclusions

The computational fluid dynamics (CFD) method was used to simulate the aerodynamic performance of a fixed wind turbine with different tilt angles. By comparing the aerodynamic performance of a wind turbine at different tilt angles, it was found the aerodynamic performance of the wind turbine is better when the tilt angle is about 4°. The main purpose of the paper was to study the practical importance of effect of tilt angle on the aerodynamic performance of a wind turbine. The main conclusions of the paper are as follows:

1. In order to balance the power generation efficiency of the wind turbine and the interaction between the blade and the tower, the tilt angle of a wind turbine can be set at about 4° to obtain better aerodynamic performance.

2. The increase of the tilt angle will cause the load of the section airfoil to change, thus affecting the thrust and power of the wind turbine. When the blade is located in front of the tower, increasing the tilt angle will increase the load of the section airfoil. At the same time, after the tilt angle reaches 8°, the change in the load of the section airfoil with the tilt angle will not be obvious.

3. Wind shear will cause the thrust and power of the wind turbine to decrease, and the effect of the wind shear exponents on the aerodynamic performance of the wind turbine is significantly greater than the expected effect of the turbulence intensity. When performing wind turbine simulations, it is recommended to use a wind shear that is closer to that found in the real environment instead of uniform wind.

In summary, in order to ensure that a fixed wind turbine has an improved aerodynamic performance, the tilt angle of the wind turbine when installed should be about 4°. In reality, for a floating offshore wind turbine, a tilt angle of about 4° may not be appropriate, so the effect of tilt angle on a floating offshore wind turbine should be further studied in future works.

Author Contributions: Conceptualization, K.L.; Data curation, Q.W.; Formal analysis, Q.W.; Software, Q.W.; Supervision, Q.M.; Validation, Q.W.; Writing—original draft, Q.W.; Writing—review & editing, K.L. and Q.M. All authors have read and agreed to the published version of the manuscript.

Funding: This work is supported by the National Natural Science Foundation of China (Nos. 51739001, 51779049).

Conflicts of Interest: The authors declare no conflict of interest.

References

1. Laks, J.H.; Pao, L.Y.; Wright, A.D. Control of wind turbines: Past, present, and future. In Proceedings of the 2009 American Control Conference, St. Louis, MO, USA, 10–12 June 2009.
2. Pao, L.Y.; Johnson, K.E. A tutorial on the dynamics and control of wind turbines and wind farms. In Proceedings of the 2009 American Control Conference, St. Louis, MO, USA, 10–12 June 2009.
3. Zhao, Q.; AlKhalifin, Y.; Li, X.; Sheng, C.; Afjeh, A. Comparative study of yaw and nacelle tilt control strategies for a two-bladed downwind wind turbine. *Fluid Mech. Res. Int. J.* **2018**, *2*, 85–97. [CrossRef]
4. Kim, H.; Lee, S.; Lee, S. Influence of blade-tower interaction in upwind-type horizontal axis wind turbines on aerodynamics. *J. Mech. Sci. Technol.* **2011**, *25*, 1351–1360. [CrossRef]
5. Guo, P. Influence analysis of wind shear and tower shadow on load and power based on blade element theory. In Proceedings of the 2011 Chinese Control and Decision Conference (CCDC), Mianyang, China, 23–25 May 2011.
6. Wang, Q.; Zhou, H.; Wan, D. Numerical simulation of wind turbine blade-tower interaction. *J. Mar. Sci. Appl.* **2012**, *11*, 321–327. [CrossRef]
7. Narayana, M. Gyroscopic Effect of Small Scale Tilt Up Horizontal Axis Wind Turbine. In *World Renewable Energy Congress VI 2000*; Elsevier: Amsterdam, The Netherlands, 2000; pp. 2312–2315. [CrossRef]
8. Zhao, Q.; Sheng, C.; Al-Khalifin, Y.; Afjeh, A. Aeromechanical analysis of two-bladed downwind turbine using a nacelle tilt control. In Proceedings of the ASME 2017 Fluid Division Summer Meeting, Waikoloa, HI, USA, 30 July–3 August 2017.
9. Abdulqadir, S.A.; Iacovides, H.; Nasser, A. The physical modelling and aerodynamics of turbulent flows around horizontal axis wind turbines. *Energy* **2017**, *119*, 767–799. [CrossRef]
10. Menter, F.R. Two-equation eddy-viscosity turbulence models for engineering applications. *AIAA J.* **1994**, *32*, 1598–1605. [CrossRef]
11. Vermeer, N.-J.; Sørensen, J.N.; Crespo, A. Wind turbine wake aerodynamics. *Prog. Aerosp. Sci.* **2003**, *39*, 467–510. [CrossRef]
12. Parente, A.; Gorle, C.; Beeck, J.v.; Benocci, C. Improved k-e model and wall function formulation for the RANS simulation of ABL flows. *Wind Eng. Ind. Aerodyn.* **2011**, *99*, 267–278. [CrossRef]
13. Tran, T.T.; Kim, D.-H. Fully coupled aero-hydrodynamic analysis of a semi-submersible FOWT using a dynamic fluid body interaction approach. *Renew. Energy* **2016**, *92*, 244–261. [CrossRef]
14. Politecnico di Milano. Qualification of innovative floating substructures for 10MW wind turbines and water depths greater than 50m. 2016. Available online: https://lifes50plus.eu/wp-content/uploads/2015/11/GA-640741_LIFES50_D3.2-mn-fix-1.pdf (accessed on 15 April 2018).
15. Miao, W.; Li, C.; Pavesi, G.; Yang, J.; Xie, X. Investigation of wake characteristics of a yawed HAWT and its impacts on the inline downstream wind turbine using unsteady CFD. *J. Wind. Eng. Ind. Aerodyn.* **2017**, *168*, 60–71. [CrossRef]
16. Jonkman, J.; Butterfield, S.; Musial, W.; Scott, G. *Definition of a 5-MW Reference Wind Turbine for Offshore System Development*; Report No. NREL/TP-500-38060; National Renewable Energy Laboratory: Golden, CO, USA, 2009.
17. Zhong, W.; Shen, W.; Wang, T.; Li, Y. A tip loss correction model for wind turbine aerodynamic performance prediction. *Renew. Energy* **2020**, *147*, 223–238. [CrossRef]

18. Yu, Z.; Zheng, X.; Ma, Q. Study on Actuator Line Modeling of Two NREL 5-MW Wind Turbine Wakes. *Appl. Sci.* **2018**, *8*, 434. [CrossRef]
19. Quallen, S.; Xing, T. An Investigation of the Blade Tower Interaction of a Floating Offshore Wind Turbine. In Proceedings of the 25th International Ocean and Polar Engineering Conference, Kona, HI, USA, 21–26 June 2015.
20. Rahimi, H.; Schepers, J.; Shen, W.; García, N.R.; Schneider, M.; Micallef, D.; Ferreira, C.S.; Jost, E.; Klein, L.; Herráez, I. Evaluation of different methods for determining the angle of attack on wind turbine blades with CFD results under axial inflow conditions. *Renew. Energy* **2018**, *125*, 866–876. [CrossRef]
21. IEC. *IEC61400-1, Wind Turbines-Part1: Design Requirements*; International Electrotechnical Commission: Geneva, Switzerland, 2005.

 *applied sciences*

Article

Experimental Study on Aerodynamic Characteristics of a Gurney Flap on a Wind Turbine Airfoil under High Turbulent Flow Condition

Junwei Yang [1,2,3], **Hua Yang** [1,2,*], **Weijun Zhu** [1,2], **Nailu Li** [1,2] **and Yiping Yuan** [4]

[1] College of Electrical, Energy and Power Engineering, Yangzhou University, Yangzhou 225127, China; yangjunwei@yzu.edu.cn (J.Y.); wjzhu@yzu.edu.cn (W.Z.); nlli@yzu.edu.cn (N.L.)

[2] New Energy Research Center, Yangzhou University, Yangzhou 225009, China

[3] College of Hydraulic Science and Engineering, Yangzhou University, Yangzhou 225009, China

[4] Jiangsu Key Laboratory of Hi-Tech Research for Wind Turbine Design, Nanjing University of Aeronautics and Astronautics, Nanjing 210016, China; yuanyiping@nuaa.edu.cn

* Correspondence: yanghua@yzu.edu.cn; Tel.: +86-138-1583-8009

Received: 20 September 2020; Accepted: 14 October 2020; Published: 16 October 2020

Abstract: The objective of the current work is to experimentally investigate the effect of turbulent flow on an airfoil with a Gurney flap. The wind tunnel experiments were performed for the DTU-LN221 airfoil under different turbulence level (T.I. of 0.2%, 10.5% and 19.0%) and various flap configurations. The height of the Gurney flaps varies from 1% to 2% of the chord length; the thickness of the Gurney flaps varies from 0.25% to 0.75% of the chord length. The Gurney flap was vertical fixed on the pressure side of the airfoil at nearly 100% measured from the leading edge. By replacing the turbulence grille in the wind tunnel, measured data indicated a stall delay phenomenon while increasing the inflow turbulence level. By further changing the height and the thickness of the Gurney flap, it was found that the height of the Gurney flap is a very important parameter whereas the thickness parameter has little influence. Besides, velocity in the near wake zone was measured by hot-wire anemometry, showing the mechanisms of lift enhancement. The results demonstrate that under low turbulent inflow condition, the maximum lift coefficient of the airfoil with flaps increased by 8.47% to 13.50% (i.e., thickness of 0.75%), and the Gurney flap became less effective after stall angle. The Gurney flap with different heights increased the lift-to-drag ratio from 2.74% to 14.35% under 10.5% of turbulence intensity (i.e., thickness of 0.75%). However, under much a larger turbulence environment (19.0%), the benefit to the aerodynamic performance was negligible.

Keywords: wind tunnel experiment; wind turbine airfoil; turbulence; Gurney flap; aerodynamic characteristics

---

1. Introduction

Wind power generation technology has been maturely developed in the past decades. Airfoil is a basic element of a wind turbine blade, and its aerodynamic characteristics have a major influence on the wind energy conversion efficiency. Among the conventional rotor aerodynamic design strategy, the blade add-ons were of particular interest to further improve wind energy efficiency. Therefore, mounting flap to the airfoil trailing edge was one of the most feasible methods to improve the aerodynamic performance of wind turbines. In addition to power production, such a technique can also effectively reduce the aerodynamic loads both of wind turbine blades and tower. If a sophisticated controller was implemented, the flap can further reduce turbulence-induced fatigue loads, so that longer lifetime was guaranteed. After the pioneering work of Liebeck [1], a large number of studies have been conducted to explain the phenomena induced by the presence of this device. More recently, research objects were most focused on airfoil attached various shapes of flaps, such as Gurney flaps [2,3], triangular flaps [4], separate trailing edge flaps [5,6] and deformable trailing edge flaps [7,8]. For experiment tests, Zhang et al. [9] and Amini et al. [10] studied the aerodynamic effect of a Gurney flap based on the airfoil through wind tunnel experiments. T Lee et al. [11] carried out a wind tunnel test of the lift force and pitching moment coefficients of both trailing edge flaps and Gurney flaps with different shape parameters. Based on a 5 MW reference wind turbine, Chen et al. [12] designed and optimized a trailing edge flaps such that the blade mass can be further reduced but still maintain the desired power performance. Medina et al. [13] explored the flow mechanisms of a flap at a high deflection angle. When the flap works in a separated flow region, he provided some ideas for realizing instantaneous action or alleviating extra aerodynamic loads on wind turbines. Elsayed et al. [14] studied the flap tip vortexes and characterized the flow structures behind a flap in a low-speed wind tunnel by using particle image velocimetry. Little et al. [15] designed a trailing edge flap by using a single medium plasma driver which resulted in a higher lift force. Bergami et al. [16] designed an active controller of a trailing edge flap on a 5 MW reference wind turbine and proved that the flap could effectively control aerodynamic loads. Edward et al. [17] also conducted experiments on a flaps noise drop. According to the wind tunnel tests based on a full-size rotor, Straub et al. [18] found that flaps could also be used to control noise and vibration such that the noise generated by blade vortex interaction could be reduced by 6 dB. For numerical simulations and theoretical analyses, Traub et al. [19] fitted a semi-empirical equation by summarizing the performance of a large number of flaps. Lario et al. [20] numerically solved the unsteady flow field of Gurney flaps at high Reynolds numbers through the discontinuous Galerkin method. By analyzing dynamic characteristics of an airfoil with Gurney flaps through the numerical simulation, Li et al. [21] found that flaps were capable to reduce unsteady aerodynamic loads of wind turbines. Zhu et al. [22] numerically simulated the airfoil with trailing edge flaps using the immersed boundary method and found that flaps could be combined with the paddle movement to adjust the aerodynamic loads of a wind turbine airfoil. Ng et al. [23] investigated the trailing edge flap together with an aeroelastic analysis. It was noted that the trailing edge flap could be a smart device to control aeroelastic deformation.

All the above researches on the flaps were built on the uniform inflow of low turbulence intensity. In the past, there were few studies carried out by flow over flaps under high turbulence intensity [24,25], however, such a flow condition often occurs on wind turbines operating in a wind farm. Considering wind turbines operate in a turbulence environment, the conclusions obtained from the previous studies might not be accurate. Therefore, to simulate wind turbines under turbulence environment, active and passive wind tunnel turbulence generation methods can be used. The active technology includes a vibrating grille and multi-fan wind tunnel; the maximum turbulence intensity can reach more than 20% [26]. The passive control structure was relatively simple, which can be divided into grille, wedge, and rough square types among the passive turbulence generators, it was more convenient to construct grilles, which have gain very popular use. In this experiment, specific turbulence levels were passively controlled by a grille with proper grid size.

The investigations presented in this paper were focused on the coupled effects of Gurney flap and turbulence inflow. The Gurney flaps were experimentally investigated under various turbulence intensities and flap configurations. The desired turbulent flow passes the airfoil was achieved by changing the grille size as well as the distance between the grille and the airfoil model. The hot-wire anemometer was used to record the wind speed and turbulence intensity in a flow cross-section. The quantitative information obtained during the experiment includes: (1) turbulent field descriptions, (2) airfoil pressure coefficients, (3) lift-to-drag coefficients, (4) wake measurements. On that basis, the aerodynamic performance of Gurney flaps with different heights and thicknesses was researched. The values of lift and pitching-moment coefficients were obtained through the integration of surface pressures. The wake rake array was also used to determine the values of drag coefficient. Furthermore, the flow fields near the trailing edge of airfoil were tested which could further verify the reason for lift improvement. Finally, concluding remarks accompany the discussion of the experimental investigations.

2. Experimental Setup

The experiments were carried out in a wind tunnel located at Yangzhou University. It was a close-loop type wind tunnel which contains two experimental sections. The measurements were conducted in the smaller section with the cross-section parameters of 3 m × 1.5 m, and the length is 3 m. The operational wind speed range was 0~50 m/s and the calibrated maximum turbulence intensity in the free stream was 0.2%. The DTU-LN221 airfoil model [27], as shown in Figure 1, was adopted which has a chord length of 0.6 m and a span length of 1.5 m. The airfoil model was vertically placed in the test section. The bottom part of the airfoil section was connected by the rotating shaft, which was fixed on a rotational plate, as demonstrated in Figure 2. Therefore, the angle of attack can be remotely controlled via a shaft connection with a motor below the wind tunnel. The airfoil model was made of aluminum alloy where small taps were drilled on the surface of the middle section, with the location sketched in Figure 1. The arc length between the taps was approximately spaced every 2.5% of chord length starting from the leading edge to 90% of the trailing edge. The hollow plastic hoses were connected on the reverse side of the airfoil surface which has an outer diameter of 1.2 mm. Thus, a total of 77 pressure taps were arranged for the pressure measurement on the surface of the airfoil. The pressures on the airfoil surface were sensed by the electronic pressure acquisition system of PSI (Pressure Systems Inc., Hampton, VA, USA). The pressure system used in the experiment has a sampling frequency of 333.3 Hz, a measuring range of ±2.5 kPa and a measuring accuracy of ±0.05%. Figure 1 also shows the chord-wise position, so the lift and pitching-moment of the test airfoil were achieved by integrating the pressure and the position of the taps on the surface. The lift coefficient and pitching-moment coefficients were given by

$$C_l = C_n \cos \alpha - C_t \sin \alpha$$
$$C_m = \int_0^1 \left(C_{pl} - C_{pu}\right)(0.25 - \hat{x})d\hat{x} \tag{1}$$

where

$$C_n = \int_0^1 \left(C_{pl} - C_{pu}\right)d\hat{x}$$
$$C_t = \int_{\hat{y}_{l\ max}}^{\hat{y}_{u\ max}} \left(C_{p,be} - C_{p,af}\right)d\hat{y} \tag{2}$$

where C_l and C_m are the lift coefficient and pitching-moment coefficient, respectively; α represents the angle of attack; $\hat{x}$ is the relative chord length; $\hat{y}$ is the thickness value relative to the chord length; C_{pu} and C_{pl} are the pressure coefficient on the suction and pressure sides of the airfoil, respectively; C_n and C_t are the normal force coefficient and tangential force coefficient, respectively. $C_{p,be}$ and $C_{p,af}$ are the pressure coefficient before and after the maximum thickness of airfoil; $\hat{y}_{u\ max}$ and $\hat{y}_{l\ max}$ represent the maximum thickness values of the suction and pressure sides relative to chord length.

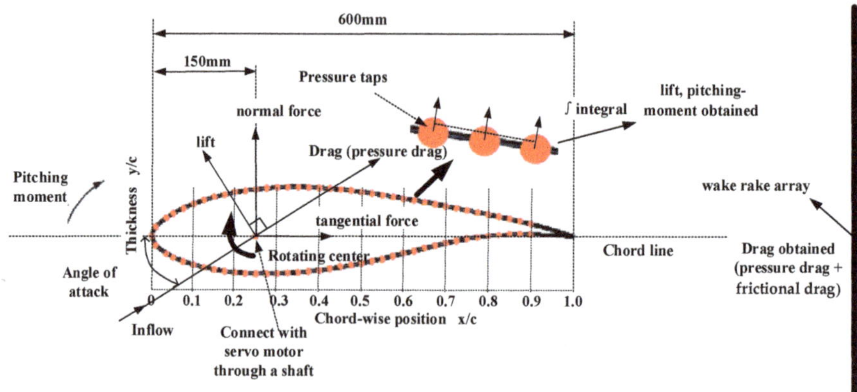

Figure 1. Schematic diagram of measuring position.

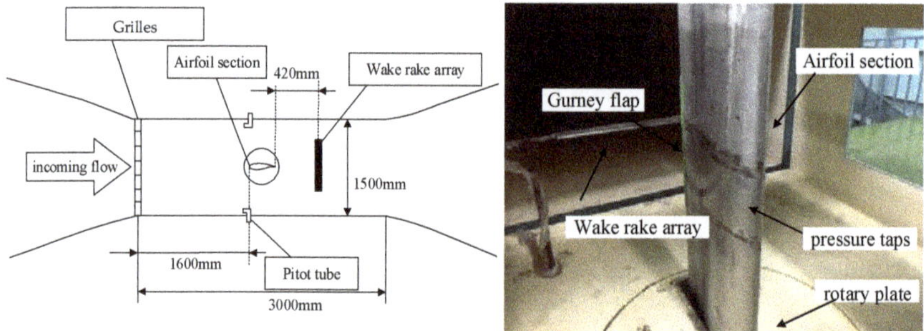

Figure 2. Experimental test device.

At a small angle of attack, the aerodynamic drag largely consists of frictional drag, while the data measured by surface pressure taps cannot accurately represent the drag force, so the drag can be measured by the momentum method more precisely. As shown in Figure 2, a wake rake array was placed at 0.7 chord length behind the trailing edge of the airfoil and at the same vertical level as the pressure taps. The measurement range of the wake probes was 80.8 cm, 102 total pressure pipes (with an outer diameter of 1.2 mm) and 4 static pressure pipes (with an outer diameter of 2 mm) were averagely arranged, with an 20 cm apart for the static pressure pipes. To prevent air leakage, all pressure tubes were connected by plastic hoses to the pressure measurement device. Besides, two Pitot tubes were installed at 1.6 m measured from the downstream of grilles where the free stream velocity was recorded. The drag coefficients were given as follows

$$C_d = \frac{2}{c} \int_w \sqrt{\frac{P_{01} - P}{P_0 - P_\infty}} \left[1 - \sqrt{\frac{P_{01} - P_\infty}{P_0 - P_\infty}} \right] ds \tag{3}$$

where C_d is the drag coefficient; c is the chord length, w is the range of integration; s is the coordinate along the thickness direction of the wake rake array; P_∞ and P_0 are the static pressure and total pressure measured by the Pitot tubes; P and P_{01} are the static pressure and total pressure measured by the wake rake array. It should be noted that there is a total pressure loss along the Pitot tubes to the wake rake array, therefore the total pressure loss should be added to each total pressure measuring point of the wake rake array.

Figure 3 presents the schematic diagram of the grille geometry. The grilles assembled with many squared alloys with geometry specified by four parameters a, b, c and d. These small squares were bolted together and on top of the grille the rubber pads were attached. During the measurements, two types of grilles were implemented. The dimensions of the two grilles were provided in Table 1. The width of the longitudinal grille and the transverse grille were denoted by a and c, respectively; the width and the height of the spacing were denoted by b and d, respectively, the thickness along the flow direction was e.

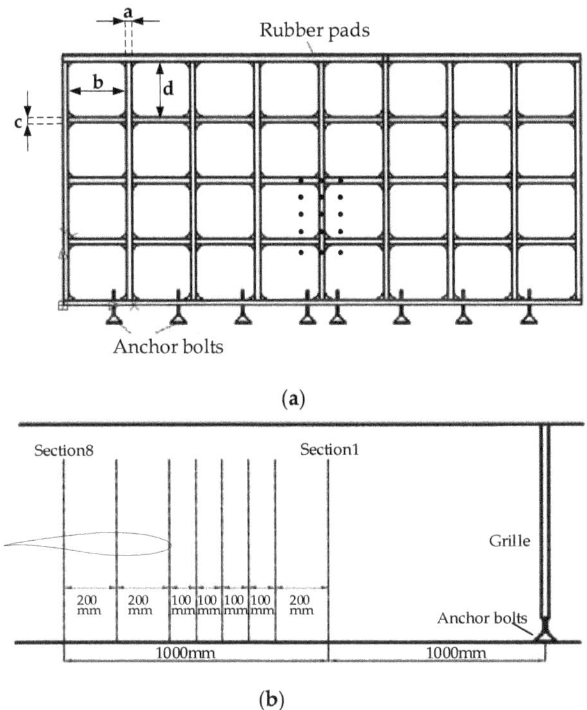

Figure 3. Schematic diagram of experimental grille. (**a**) Front view; (**b**) side view.

Table 1. Dimensions of the experiment grilles.

	a (cm)	b (cm)	c (cm)	d (cm)	e (cm)
Scheme 1	3	34	3	32	3
Scheme 2	6	31.4	6	29.8	3

Behind the grille, there were eight measurement sections designed for velocity recording. At each of the section, 15 measurement points were uniformly spaced. The center of the topmost measurement point corresponds to the geometrical center of the wind tunnel section. The horizontal and vertical spacing of measuring points were 10 cm, as displayed by the black spots in Figure 3a. In addition, we set eight measurement sections one to two meters behind the grille, as shown in Figure 3b. The wind speed was measured through a Dantec hot-wire probe (type 55P61). The sampling frequency was 5 kHz, and the test wind speeds were consistent with that in the airfoil experiment. Figure 4 displays the device of turbulence measurement. Besides, wake measurements in this paper were carried out on an electronically controlled traverser consisting of a two-dimensional probe, as displayed in Figure 5.

Figure 4. Grille turbulence generated device.

Figure 5. Wake measurements.

The influence of the flap height and thickness were two key parameters which influence the overall aerodynamic performance of the airfoil. For this reason, a total of nine types of flaps were designed in the experiments, namely, the flap with heights h of 1%, 1.5% and 2% chord length (6 mm, 9 mm and 12 mm, respectively) and with relative thicknesses d of 0.25%, 0.5% and 0.75% chord length (1.5 mm, 3 mm and 4.5 mm, respectively). The diagram of the flap used in the experiment was given in Figure 6. As shown, the flap has an L shape; the bottom part has a width of 7 mm which was attached to the pressure side of the airfoil trailing edge. During the experiments, the flap was vertically attached near the position of 100% of the chord length, see Figure 2 with the specific experimental setup. The coordinate system behind the trailing edge was defined for subsequent wake measurements, also the x and y axis were set in the vertical and the forward directions, respectively. The black spots in the Figure 6 represented the hot-wire measuring positions set at downstream of the trailing edge.

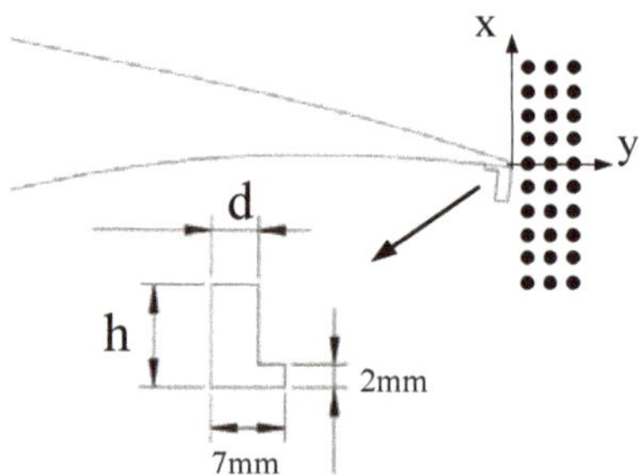

Figure 6. Flap geometry.

3. Experimental Results and Analysis

In the experiment, the free stream flow velocity was 20 m/s, and the corresponding Reynolds number based on the airfoil chord length was 0.8×10^6, although wind turbines often operate at wind speed below 20 m/s, but the magnitude of the Reynolds number was up to an order of 10^6. Therefore, the experimental value of the Reynolds number was chosen to approach an order of magnitude corresponding to those obtained from full-scale wind turbines. Two types of grilles generate different turbulence levels in the wind tunnel, such that the experiments were mainly separated into low and high turbulence cases. To be specific, a grille was placed at the upstream of the airfoil test section which results in different aerodynamic characteristics of flow over the airfoil with and without a Gurney flap.

3.1. Experimental Results of Grille Turbulence Generator

The measurements were first conducted for the two grilles. Figure 7 illustrates the time series of the instantaneous velocity at the central measuring point of the 6th section. As shown in Figure 7, the incoming wind became highly disturbed, for this reason, a statistical analysis was done for the flow field. The decay of turbulence intensity from the grilles towards the airfoil test section was reported in Figure 8a. Within the scope of the test, the averaged turbulence intensity was recorded along 8 downstream sections (each measured with 15 points). Besides, the margin of error in Figure 8a was expressed as the standard deviation of each section data. According to Figure 8a, the average turbulence intensity of downstream in scheme one dropped from 13.5% at the position of 1.0 m behind the grille to 9.4% at the position of 2.0 m; the average turbulence intensity of the downstream in scheme two was higher than scheme 1, which decreased from 27.7% at the position of 1.0 m behind the grille to 15.9% at the position of 2.0 m. The standard deviations of turbulence intensity of the two schemes drops quickly towards the test section. The change in turbulence level indicates that there were very high turbulence fluctuations just behind the grille which yielded very unstable flow field. The flow became gradually stabilized owing to the mutual dissipation over a propagation distance. To ensure a homogeneous and isotropic turbulent field, so the distance 1.6 m (from the leading edge of the airfoil to the grille) was selected for placing the leading edge of the airfoil. The turbulence intensity measured at the 6th section of the two schemes was 10.5% and 19.0%, respectively. Figure 8b illustrates the power spectral density (PSD) of the wind speed at the two grille schemes measured at the central measuring point of the 6th section (1.6 m measured from grille to the hot-wire probes). Obviously, the larger the turbulence intensity was generated, the larger the amplitude of the power spectral density was observed. Moreover, the power spectra amplitude in the downstream direction was consistent with the corresponding Karman spectra, and it dropped down rapidly after the frequency over 30 Hz, which suggests that the grille distance was more proportional to the turbulent kinetic energy.

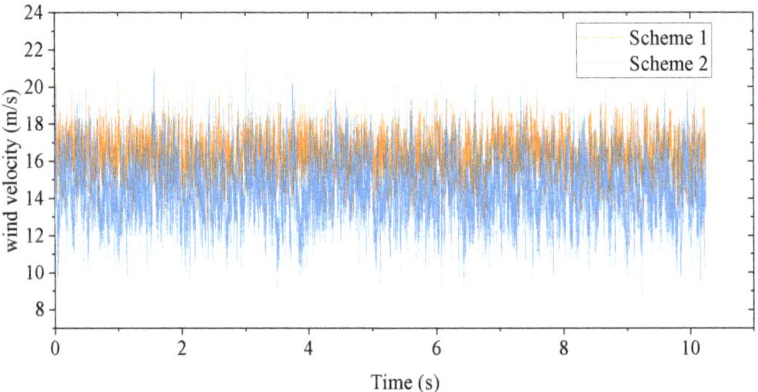

Figure 7. The time series of the instantaneous streamwise velocity (the 6th section, central measuring point).

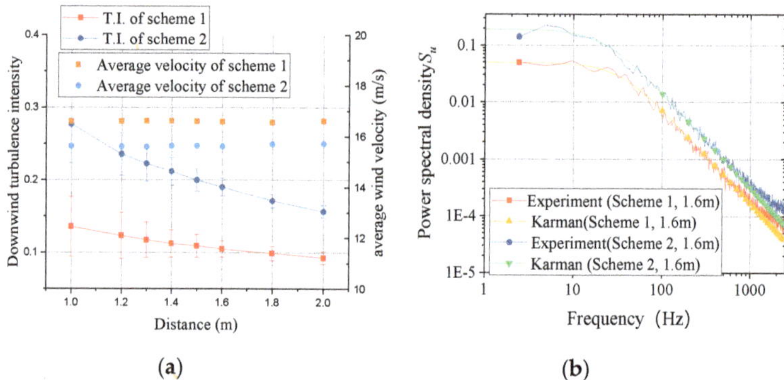

Figure 8. Description of the flow field behind the grilles. (**a**) Turbulence intensity and average velocity (Re: 0.8×10^6); (**b**) Comparison of the PSD (1.6 m behind the grille).

3.2. Experimental Results of the Baseline Airfoil

To verify the accuracy of the experiment, the experimental data from the wind tunnel of LM Wind Power were selected as a reference [28]. The aerodynamic data of the DTU-LN221 airfoil under uniform inflow were measured in the LM wind tunnel experiment at the approximate Reynolds number of 1.5×10^6. The lift and drag data were selected as a cross-validation case. In consideration of the interference effect of wind tunnel wall, the maximum blockage ratio in this experiment was about 8.4%, so the experiment measured data of the airfoil were corrected by after the reference [29], and the experimental data analyzed below were all corrected. In order to perform a comparison, Figure 9 shows the comparison of experimental data of the DTU-LN221 airfoil under uniform inflow condition. To minimize the uncertainty, measurements were performed several times for comparison test at the approximate Reynolds number of 1.5×10^6, YZU1, YZU2, YZU3 represents the results of the repeated experiments. As shown in Figure 9, the lift and drag coefficients were very similar before the stall angle, while there were some discrepancies in the stall state. Consider the cause of flow separation, the deviation of results of drag is quite considerable, but the stall angles tested in the two wind tunnels were nearly the same. The reason for this may be that although the airfoils were of the same type, they were individually manufactured, for example the roughness of the airfoil surface may affect the experimental results. Besides, there were some differences in drag measurements, LM adopted the model surface pressure distribution in a certain range of the angle of attack, while YZU used the wake probes at all angles. Similar phenomena were seen in different wind tunnel laboratories [30–32]. In the following, for reason of comparisons, the measured data are averaged.

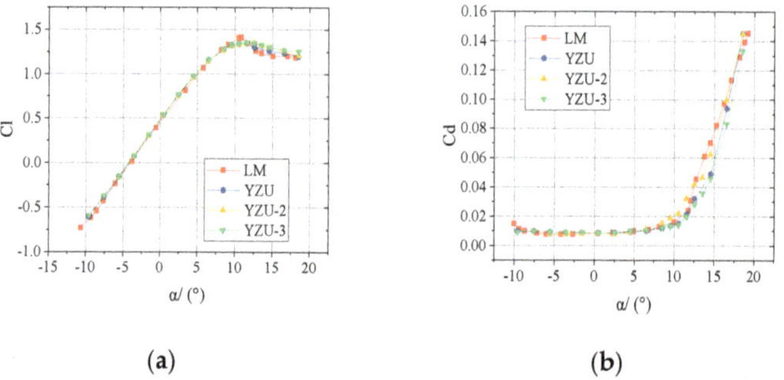

Figure 9. Comparison of experimental data of the DTU-LN221 airfoil (uniform inflow, Re: 1.5×10^6). (**a**) Lift coefficient comparison; (**b**) drag coefficient comparison.

3.3. Experimental Results of the Gurney Flap under Uniform Inflow

This subsection starts to investigate the aerodynamic characteristics of the effect of Gurney flap in low turbulent inflow condition, the Reynolds number was decreased to 0.8×10^6. Seen from the Figure 10, three lift and drag curves were compared with the original at the angle of attack ranged from $-9.6°$ to $14.4°$. As shown, the lift coefficients increased with the increase of the height of the Gurney flap before the stall angle, while it was also realized that the effect of the Gurney flap was limited at a high angle of attack. The lift coefficient slopes and the stall angles of the flapped airfoil remain essentially unchanged compared to the baseline airfoil, but the flapped airfoil caused a leftward shift compared to the baseline airfoil. In order to show the changes of aerodynamic characteristics clearly, the following lift and drag differences were defined:

$$\Delta cl = \frac{\left|cl_{bs} - cl_{gf}\right|}{cl_{bs}} \times 100\% \tag{4}$$

$$\Delta cd = \frac{\left|cd_{bs} - cd_{gf}\right|}{cd_{bs}} \times 100\% \tag{5}$$

$$\Delta cld = \frac{\left|cld_{bs} - cld_{gf}\right|}{cld_{bs}} \times 100\% \tag{6}$$

where cl_{bs} and cd_{bs} are the lift and drag coefficients obtained from the baseline airfoil, respectively. cl_{gf} and cd_{gf} are the lift and drag coefficients obtained from the flapped airfoil. cld_{bs} and cld_{bs} are the lift-to-drag ratio obtained from the baseline airfoil and flapped airfoil.

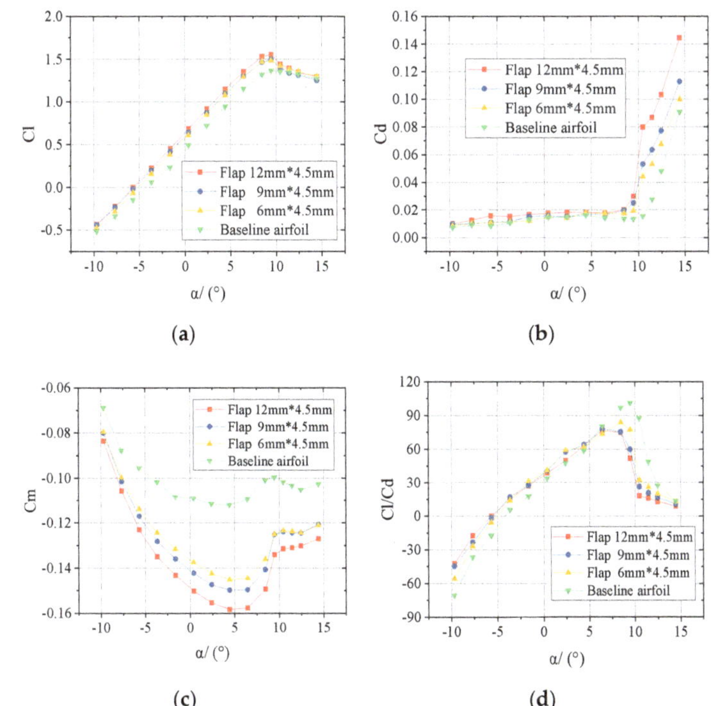

Figure 10. Variation of the aerodynamic characteristics with different heights of Gurney flaps (uniform flow, Re: 0.8×10^6). (**a**) Lift coefficient comparison; (**b**) drag coefficient comparison; (**c**) pitching-moment coefficient comparison; (**d**) lift-to-drag ratio comparison.

When the flap height changes to 6 mm, 9 mm and 12 mm, the maximum lift coefficients were increased by 8.47%, 9.56% and 13.50% at 9.4 degrees, respectively. Looking into the drag coefficients shown in Figure 10b, within the range of −9.6~10.4°, the drag coefficients show regular change where the frictional drag dominates as expected. The drag coefficients rise rapidly in the stall state, at this time the change in drag coefficient was dominated by pressure drag. For example, at an angle of attack of 12.4°, with the flap heights of 6 mm, 9 mm, 12 mm the drag coefficients were increased by 39.67%, 59.92% and 113.43%, respectively. According to Figure 10c, the Gurney flap generated a prominent increase in the pitching-moment coefficient compared to the baseline airfoil, the value increased with the heights of the Gurney flaps and reached negative peak values at approximately 4.4°. Based on Figure 10d, the lift-to-drag ratios were compared. When the angle of attack was less than 8.4°, the lift-to-drag ratios were all larger than the baseline airfoil. However, the maximum lift-to-drag ratios of the Gurney flap at three heights were smaller than the baseline airfoil. When the flap heights were 6 mm, 9 mm and 12 mm, the corresponding maximum lift-to-drag ratios were decreased by 17.16%, 22.79% and 24.47%, respectively. In the range after the stall angle, the presence of the Gurney flap reduces the lift-to-drag ratios and the higher the flap height, the more the lift efficiency decreases. When the angle of attack was 12.4°, for example, the lift-to-drag ratios of the flap with heights of 6 mm, 9 mm and 12 mm were decreased by 26.55%, 38.11% and 52.20%, respectively. Meanwhile, we observed the same trend from the other thicknesses of the Gurney flaps (i.e., of 1.5 mm), When the flap height changes to 6 mm, 9 mm and 12 mm, the maximum lift coefficients were increased by 9.27%, 9.78% and 14.08% at 9.4 degrees.

The following study in this section will be focused on the effect of the thickness of the Gurney flap. Figure 11 illustrates the changes in the aerodynamic characteristics of the baseline airfoil and airfoil with a height of 9 mm of Gurney flaps under uniform flow. As seen from the figure, by changing the thickness of the Gurney flap, the lift-to-drag curves of the flapped airfoils with different thicknesses almost overlaps, it seems that the flap thickness does not have obvious effect in the aerodynamic performance. The results imply that the small increase of the flap thickness does not change the degree of downward turning of the mean flow and recirculation flow around the trailing edge of the airfoil.

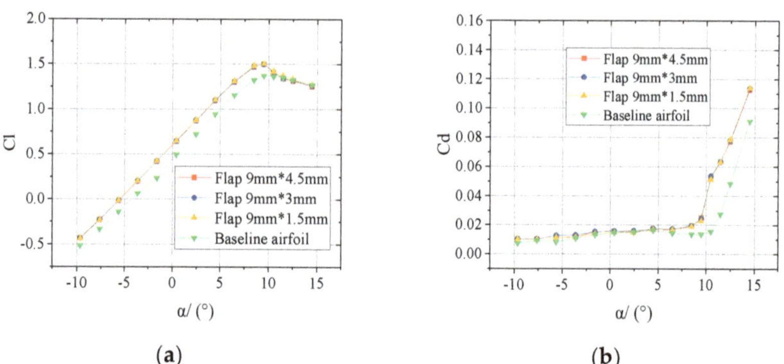

(a) (b)

Figure 11. *Cont.*

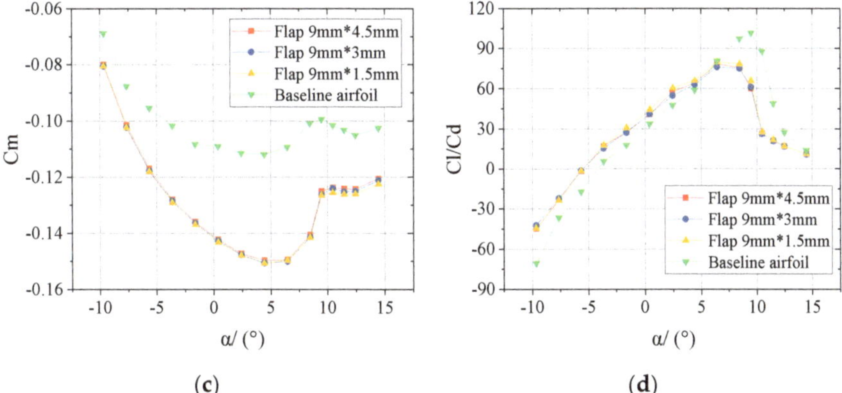

Figure 11. Variation of the aerodynamic characteristics with different thicknesses of Gurney flaps (uniform flow, Re: 0.8×10^6). (**a**) Lift coefficient comparison; (**b**) drag coefficient comparison; (**c**) pitching-moment coefficient comparison; (**d**) lift-to-drag ratio comparison.

3.4. Experimental Results of Gurney Flap under 10.5% Turbulence Intensity

This subsection starts to investigate the aerodynamic characteristics of the Gurney flap under a turbulent inflow with an intensity of 10.5%. Figure 12 shows the changes in the aerodynamic characteristics of the baseline airfoil and the Gurney flaps with a thickness of 4.5 mm when the turbulence intensity of the incoming flow was 10.5%. The measured angle of attack ranged from −9.6° to 20.4°. As shown, under the present turbulence intensity of 10.5%, all cases induced a stall delay. The stall angle was nearly delayed from the original 9.4° to about 16.4° under turbulent flow. In Figure 12a, with the increase of turbulence intensity, the maximum lift coefficient of the baseline airfoil reaches 1.611 and increases by 17.59% as compared with uniform inflow condition. An interesting observation was that in such a turbulence environment, the lift coefficients were not significantly different from each other under the current three flapped configurations. For the flap heights of 6 mm, 9 mm and 12 mm, the maximum lift coefficients were increased by 15.21%, 17.19% and 17.26% at 18.4 degrees, respectively. The increase rate was found larger than that under the uniform turbulence condition. The margin of error in Figure 12a was expressed as the variance of the time-averaged lift coefficient. As can be seen from the Figure 12a, the pressure uncertainty increased with the increase of the angle of attack, and the lift coefficient fluctuations caused by the pressure fluctuations were the most obvious pronounced when the attack angle reaches the stall angle. According to Figure 12b, the increase in the lift of the flapped airfoil was at a cost of increasing drag. The drag coefficient of the flapped airfoils was also non-linearly raised in the stall state. For example, when the angle of attack was 20.4°, the drag coefficients were increased by 12.50%, 37.50% and 44.53%, respectively. As can be seen from Figure 12c, compared with the baseline airfoil, the Gurney flap also generated a prominent increase in the pitching-moment coefficient, the negative peak values delay to about 6.4°compared with the uniform inflow condition. As demonstrated in Figure 12d, the peak value of the lift-to-drag ratio becomes less as compared with the uniform flow condition in all cases. Unlike the low turbulence condition, the maximum lift-to-drag of Gurney flap with different heights slightly increased relative to the baseline airfoil. The maximum lift-to-drag raised 14.35%, 9.15% and 2.74% respectively when the flap heights were 6 mm, 9 mm, and 12 mm, respectively, with the maximum lift-to-drag ratio occurred at the smallest flap height. Besides, in contrast to the situations of 1.5 mm thickness of the Gurney flaps, the Gurney flaps also achieved a lift increase under 10.5% turbulence intensity. The maximum lift coefficient was increased by 14.77%, 14.83%, and 16.44% for height equals to 6 mm, 9 mm and 12 mm, respectively. The two thickness configurations both had maximum lift-to-drag efficiency at 6 mm height, for the three heights, respectively.

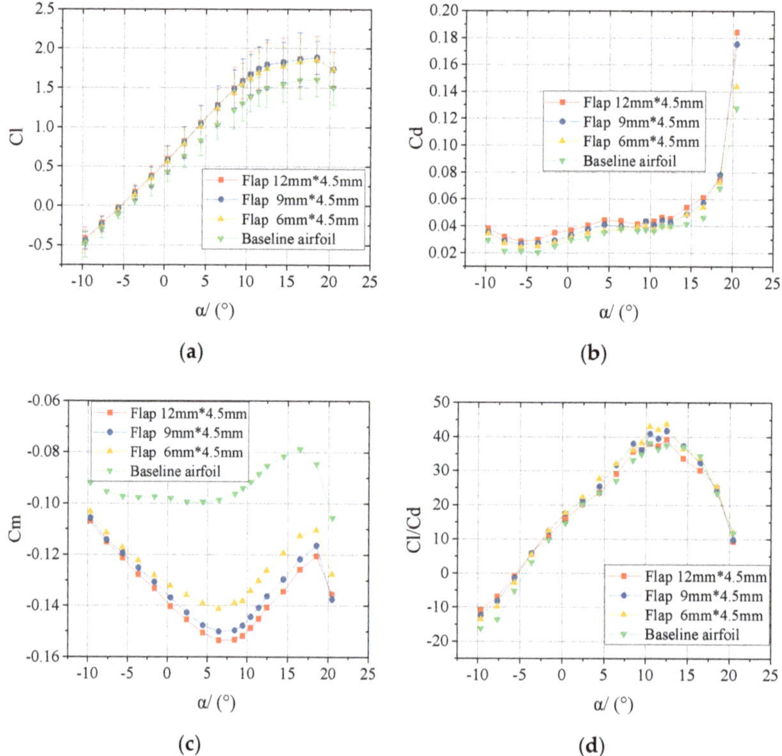

Figure 12. Variation of the aerodynamic characteristics with different heights of Gurney flaps (T.I. of 10.5%, Re: 0.8×10^6). (**a**) Lift coefficient comparison; (**b**) drag coefficient comparison; (**c**) pitching-moment coefficient comparison; (**d**) lift-to-drag ratio comparison.

In addition, the effects of various flap thicknesses were investigated. Figure 13 indicates the change of the aerodynamic characteristics of the baseline airfoil and airfoil with a height of 4.5 mm of the Gurney flaps when the turbulence intensity was 10.5%. As Figure 13 shows, like low turbulence conditions, after changing the thickness of the Gurney flap under turbulence condition of 10.5%, the aerodynamic characteristics of flapped airfoil almost show no change. The main difference in the Figure 13d was concentrated in the large lift-to-drag ratio zone, obviously because the pressure fluctuations caused by the turbulence and the flow separation at the airfoil surface result in the deviation of the measurement.

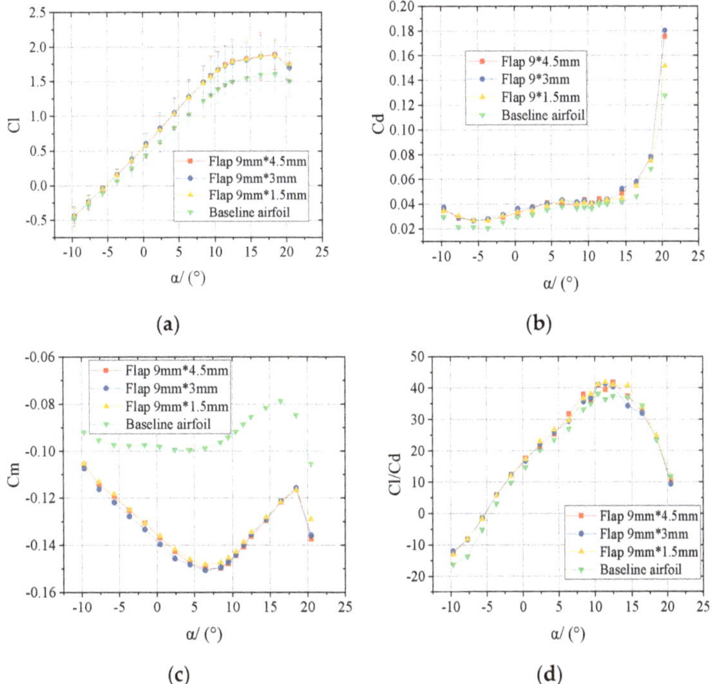

Figure 13. Variation of the aerodynamic characteristics with different thicknesses of Gurney flaps (T.I. of 10.5%, Re: 0.8×10^6). (**a**) Lift coefficient comparison; (**b**) drag coefficient comparison; (**c**) pitching-moment coefficient comparison; (**d**) lift-to-drag ratio comparison.

3.5. Experimental Results of Gurney Flap under the 19.0% Turbulence Intensity

To study the aerodynamic characteristics of the Gurney flap under high turbulence conditions, Figure 14 shows the changes of aerodynamic characteristics under the turbulence condition of 19.0%. The measured angle of attack ranged from −9.6° to 24.4°. As seen in Figure 14, the presence of turbulence enhanced the lift coefficient and drag coefficient for all cases. In Figure 14a, with the increase of turbulence intensity, the maximum lift coefficient continually increased, and the maximum lift coefficient of the baseline airfoil reaches 1.781 at 18.5 degree, with a 30.00% increase relative to that under low turbulence condition. The maximum lift coefficients increased by 2.13%, 3.37%, 5.39% at 18.5 degrees, respectively relative to the baseline airfoil, when the heights of the flaps were 6 mm, 9 mm, and 12 mm. Furthermore, since the pressure fluctuations were determined by the fluctuation's energy of the turbulent flow, the fluctuations of lift coefficient were increased relative to that under the turbulence condition of 10.5%. According to Figure 14b, the drag coefficient of the baseline airfoil changes more obvious than those under the low turbulence condition. The drag coefficients were no longer gentle within the range before stall, and it also rises sharply in the stall state. Such a trend was more noticeable after adding the Gurney flap. For example, at the angle of attack was 22.4°, the drag coefficients increased by 32.11%, 37.22% and 37.96%, when the flap heights were 6 mm, 9 mm, and 12 mm respectively. As can be seen from Figure 14c, the Gurney flap also produced a significant increase in the pitching-moment coefficient and such situation still existed in turbulent flow condition. Based on Figure 14d, unlike that under the condition of 10.5%, the lift gain obtained with Gurney flap is, unfortunately, coupled to a larger increase in drag. Adding Gurney flap with different heights at most of the angles reduces the lift-to-drag relative to the baseline airfoil, and the Gurney flap doesn't show obvious effect under the turbulence condition of 19.0%. At the same time the same situation occurred at other thicknesses of the Gurney flaps.

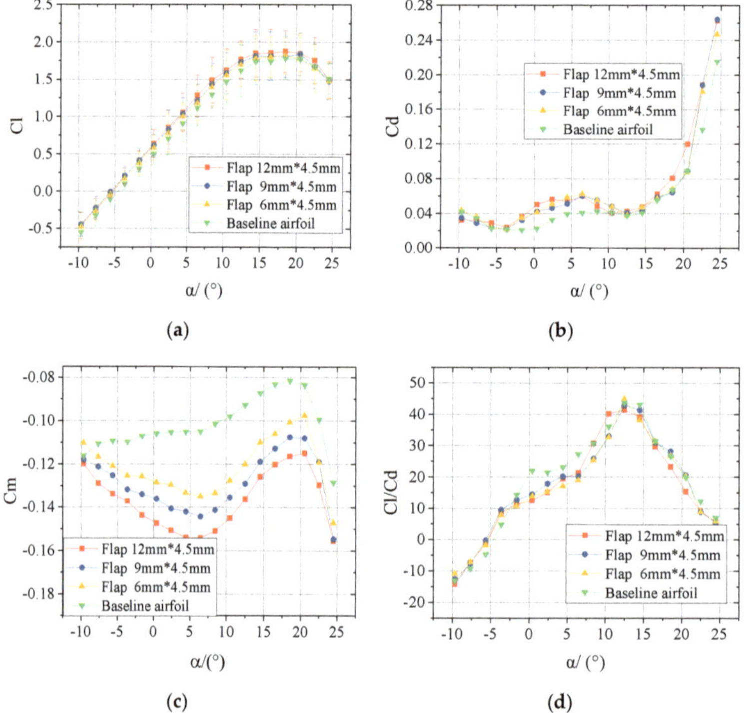

Figure 14. Variation of the aerodynamic characteristics with different heights of Gurney flaps (T.I. of 19%, Re: 0.8×10^6). (**a**) Lift coefficient comparison; (**b**) drag coefficient comparison; (**c**) pitching-moment coefficient comparison; (**d**) lift-to-drag ratio comparison.

Figure 15 exhibits the changes in the aerodynamic characteristics of the baseline airfoil and airfoil with Gurney flaps of height 4.5mm under the turbulence condition of 19.0%. As can be seen from it, with the increase of turbulence intensity, obviously, there were more pressure fluctuations. The curves of flapped airfoils with different thickness no longer overlap totally, in the majority of angles, the difference from 2% to 5%, though it may sometimes reach 20%.

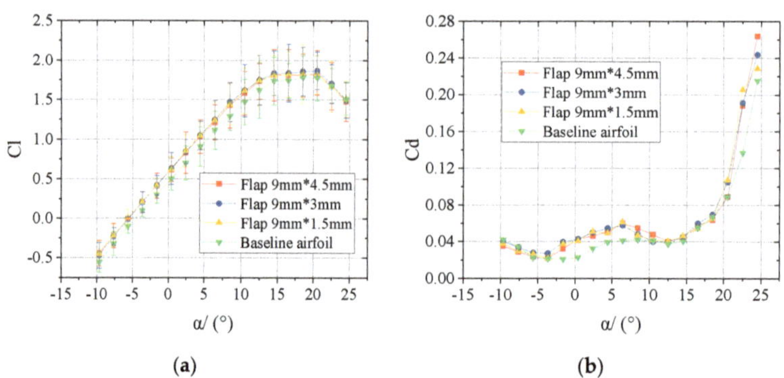

Figure 15. *Cont.*

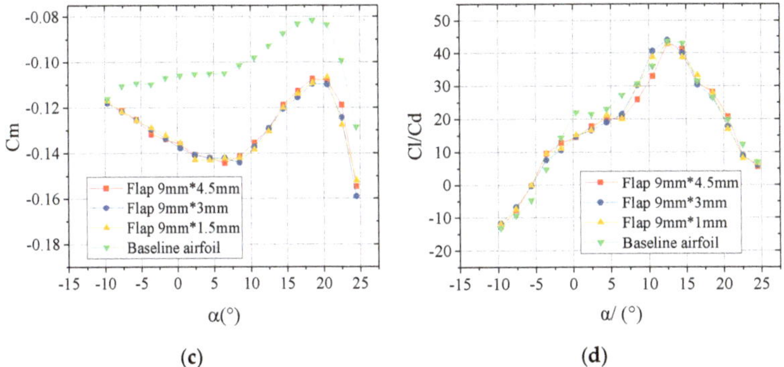

(c)

(d)

Figure 15. Variation of the aerodynamic characteristics with different thicknesses of Gurney flaps (T.I. of 19%, Re: 0.8×10^6). (**a**) Lift coefficient comparison; (**b**)drag coefficient comparison; (**c**) pitching-moment coefficient comparison; (**d**) lift-to-drag ratio comparison.

3.6. Surface Pressure Characteristics

To analyze the mechanism of turbulent inflow coupled with Gurney flap, Figure 16 illustrates the pressure coefficient distribution of the baseline airfoil and airfoil attached with the Gurney flap under different turbulent inflow when Reynolds number Re = 0.8×10^6. The x-axis shows the chord-wise position x/c and the y-axis shows the pressure coefficients C_p. The pressure coefficients C_p on the airfoil surface were given

$$C_p = \frac{p_i - p_0}{0.5\rho U_0^2} \tag{7}$$

where p_i is the pressure at the pressure tap of i (i = 1:77), p_0 is the free stream static pressure at the airfoil tested by the Pitot tube, ρ is the air density and U_0 is the free stream velocity.

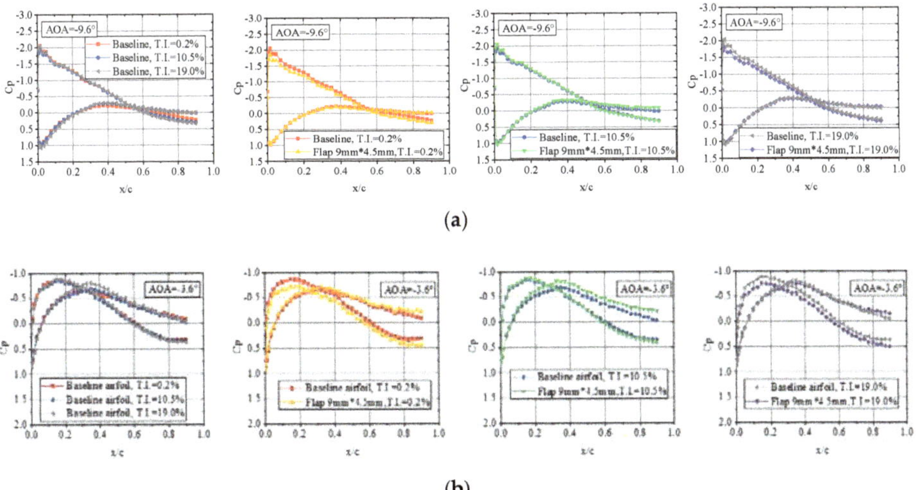

(a)

(b)

Figure 16. *Cont.*

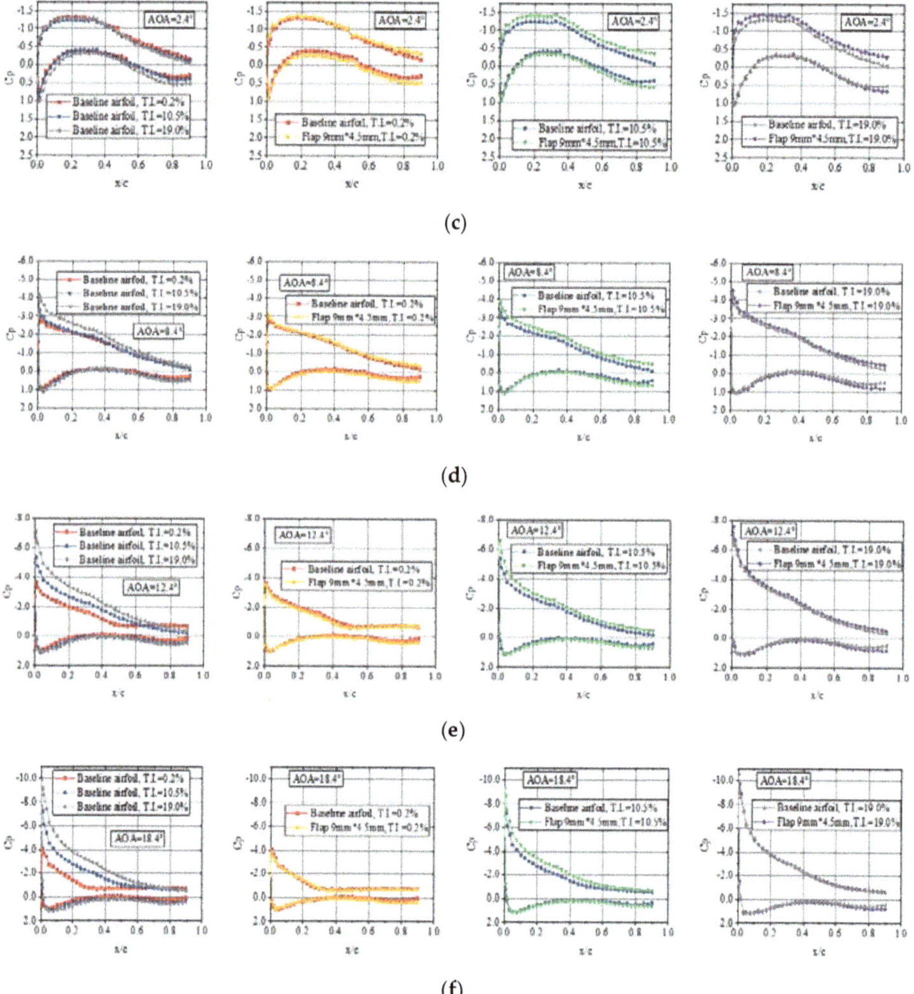

Figure 16. The surface pressure coefficients under different angle of attack (Re: 0.8×10^6). (**a**) $\alpha = -9.6°$; (**b**) $\alpha = -3.6°$; (**c**) $\alpha = 2.4°$; (**d**) $\alpha = 8.4°$; (**e**) $\alpha = 12.4°$; (**f**) $\alpha = 18.4°$.

Figure 16 illustrates that the pressure coefficients of the baseline airfoil under different turbulence conditions with small angles were similar to each other. However, with the increase of angle of attack, the differences become more pronounced at the leading edge. In contrast to the Gurney flaps, Figure 16a shows that at the angle = $-9.6°$, the differences between the flapped airfoils were small. With the increase of the angle of attack, as the angle of $-3.6°$, compared with the baseline airfoils, the intersection of the airfoil with Gurney flap moves to the leading edge under different conditions. Besides, the pressure coefficients (with the angle = $-3.6°$ and $2.4°$), the pressure distribution indicate a larger the pressure difference in the range of $0.8 < x/c < 0.9$. When the angle of attack was $2.4°$ and $8.4°$, both the pressure coefficient absolute values of the baseline airfoil and the Gurney flap tendency to increase with the increase of turbulence intensity, and the flapped airfoil becomes more obvious than the baseline airfoil. At a large angle of attack, due to the condition of stalling (Figure 16e,f), there was significant flow separation both on the suction side of flapped airfoil and baseline airfoil under uniform inflow. The larger the turbulent inflow, the larger the peak value on the leading edge of the suction side, and also the larger pressure coefficient on the pressure side.

As seen from the above, the influence of turbulent inflow coupled with Gurney flap on the pressure coefficients of airfoil was complicated. Turbulent inflow changes the flow around the airfoil surface and restrains the flow separation, after attaching the Gurney flap, the pressure distribution on both suction and pressure sides changed more significantly. Especially in the turbulence condition, the trailing edge pressure distributions were farther apart for the flapped airfoil. In these situations, the increased peak values of pressure coefficients at the leading edge increased the integral area of the pressure coefficients, so the lift force was therefore increased.

3.7. Wake Profile Characteristics

Figure 17 presents the distribution of wake velocity measured by the wake rake array under uniform inflow. As can be observed, when the angle of attack was 8.4°, which was before the stall angle of attack, the wake of airfoil with the Gurney flap was inclined to the pressure side of the airfoil. When the angle of attack was 11.4°, which was after the stall angle, the Gurney flap weaken the ability of the wake position deflection. Moreover, the wake velocity deficit was significantly increased, and the Gurney flap began to have side effects on the aerodynamic characteristics of the airfoil. The higher the Gurney flap is, the larger the wake velocity deficit is, so it indicates a higher mean drag.

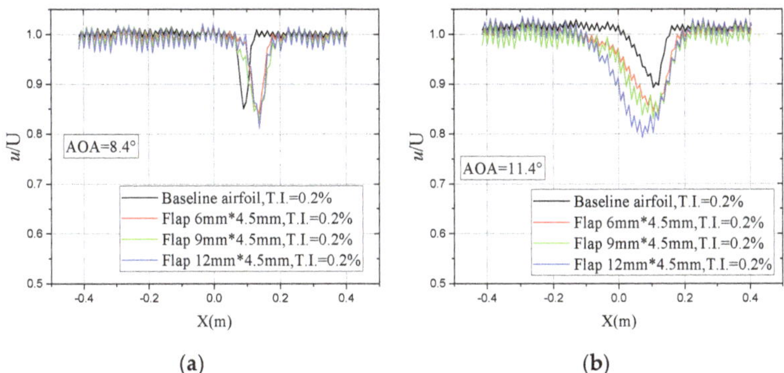

Figure 17. The wake velocity measured by the wake rake array (uniform inflow, Re: 0.8×10^6). (a) $\alpha = 8.4°$; (b) $\alpha = 11.4°$.

By comparing the flows in the near trailing edge region of the airfoil with Gurney flap. The resultant velocity distribution of the baseline airfoil and the airfoil attached with Gurney flap (9 mm in height and 4.5 mm in thickness) at the angle of attack 8.4° (before the stall angle) and 11.4° (after the stall angle) under uniform inflow were depicted in Figure 18. The vertical direction Y denoted the forward distance from the hot-wire probe to the trailing edge, and the horizontal direction X denoted the lateral distance from the hot-wire probe to the trailing edge. As shown in Figure 18, the black solid spots were the actual measurement positions by using the electronically controlled traverser. As shown, the wake flows shift to the pressure side in both the two situations. When the angle of attack was 8.4°, the airfoil with Gurney flap had a larger velocity deficit than the baseline airfoil, which suggests an increased drag. When the angle of attack was 11.4°, the velocity deficit of the baseline airfoil was also less than the Gurney flap, but there was no obvious difference between the offset direction caused by the flapped airfoil and the baseline airfoil which were being similar to the far field results measured by the wake rake array.

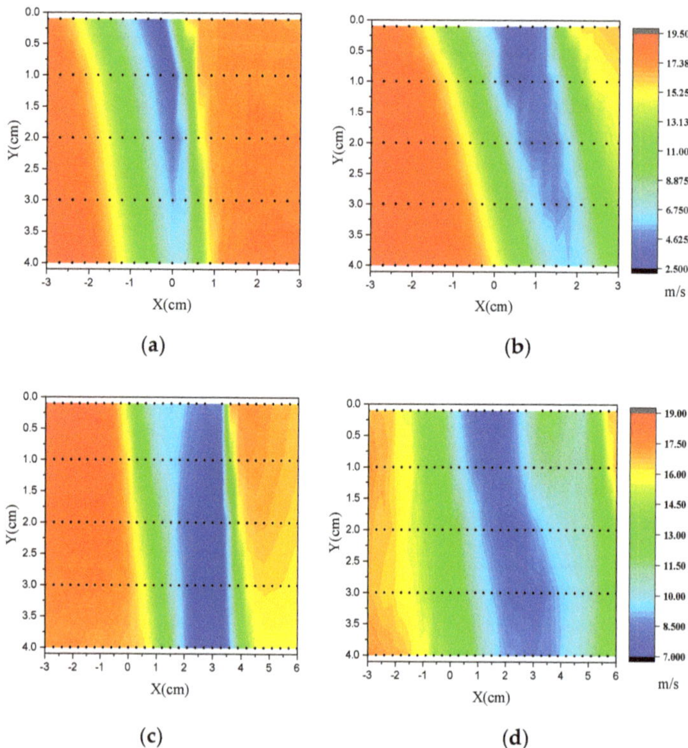

Figure 18. Velocity distribution of the region behind trailing edge (uniform inflow, Re: 0.8×10^6). (**a**) $\alpha = 8.4°$, baseline airfoil; (**b**) $\alpha = 8.4°$, Gurney flap 9 mm × 4.5 mm; (**c**) $\alpha = 11.4°$, baseline airfoil; (**d**) $\alpha = 11.4°$, Gurney flap 9 mm × 4.5 mm.

Figure 19 presents the power density spectrum measured by the region 1 cm directly behind the trailing edge under uniform inflow. As can be observed, at the angle of attack 8.4°, both along-wind and across-wind display an obvious peak in the power density spectrum, meanwhile the across-wind energy increased in the flapped airfoil. It indicates there are counter rotating vortices generated by the Gurney flap, while the wake of baseline airfoil does not show the existence of vortex shedding [33]. Gurney flaps also have a wake structure similar to the Karman vortex street behind the cylinder at bigger Reynolds number (i.e., of 0.8×10^6), the flow oscillations were caused by the separating of the shear layer on the pressure side and suction side of the trailing edge. The Karman vortex street both can be found in uniform flow and turbulent flow, so the increased lift produced by the Gurney flap seems can be explained in turbulent flow condition, either. Moreover, from the lift coefficients shown above, the size of structures was affected by turbulence level. Besides, as compared with the reference data under similar condition [33], with the increase of the Reynolds number, from Re = 1×10^5 to 0.8×10^6, the Strouhal number decreased from 0.151 (Re: 1×10^5, 2% chord length) to 0.128 (Re: 2×10^5, 2% chord length) and to 0.088 (Re: 0.8×10^6, 1.5% chord length).

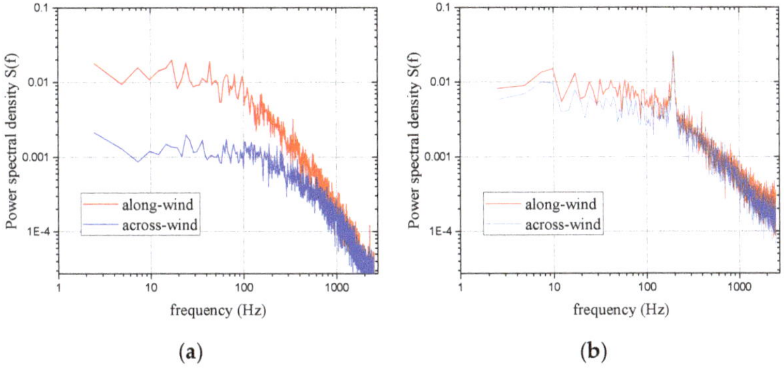

Figure 19. Comparison of the power density spectrum (uniform inflow, Re: 0.8×10^6). (**a**) $\alpha = 8.4°$, baseline airfoil; (**b**) $\alpha = 8.4°$, Gurney flap 9 mm × 4.5 mm.

4. Conclusions

In summary, this paper presented a study of the aerodynamic performance of airfoils with/without the Gurney flap under different turbulence conditions by means of wind tunnel experiments. The experimental observations were as follows:

The Gurney flap deflects the wake position from the pressure side of the airfoil, increasing the vertical distance between the airfoil chord and the middle arc. The surface pressure characteristics revealed that the pressure difference of the trailing edge was larger for the flapped airfoil in the turbulence condition. Moreover, due to the special wake structure, the Gurney flap also indicates the vortex generated at the trailing edge, which results in a greater lift of the airfoil. The lift increment obtained from Gurney flap was influenced by the turbulence intensity, meanwhile a large increase in drag was observed as well. Under a medium range of turbulence intensity (i.e., of 10.5%), the maximum lift-to-drag ratios increased from 14.35% (flap 6 mm × 4.5 mm, a = 12.4°) to 14.47% (flap 6 mm × 1.5 mm, a = 12.4°), the 1.0% Gurney flap case has the best performance, that's probably because the decrease of boundary layer thickness under turbulent conditions, so the height of 1.0% was optimal. However, under a much larger turbulence inflow (i.e., of 19.0%), all experimental data have shown that the impact was negligibly small. The drag increment was found to surpass the benefit of the increase in lift, resulting in a worse lift-to-drag ratio than the baseline airfoil.

It was of importance to note that there are two kinds of measurement uncertainties could affect the acquired results. One was stochastic uncertainties caused by the discrete pressure data which can be eliminated by multiple measurements. The other one was systematic uncertainties caused by measuring instruments which can be reduced by calibration of the instruments before the experiment.

Based on the observations from the present measurements, it was found that under very high turbulence level, e.g., with 19% of turbulence intensity, the Gurney flap has very small influence on the airfoil aerodynamic performance. The measurement results show the Gurney flap may be used in areas with slightly lower turbulence. Nevertheless, wind turbines in real life were running in the turbulent atmospheric boundary layer, as shown above, the effects of Gurney flap are highly dependent on turbulent inflow, it seems that the Gurney flap would be installed as a scalable form and preferred to have smaller size, such that it would get more potential benefits at all periods of rotation. In addition, it should be noted that the incoming flow is often a combination of rotational wake from the previous wind turbines. Research on the lift increment of Gurney flap was not isolated, it always interacted with such highly unsteady aerodynamic issues. Therefore, to analyze more precisely the effects of combining such turbulence levels with Gurney flap, further experiments should be carried out for the rotor blades in highly turbulent flows. This was in progress on a wind turbine blade composed of several DTU-LN221 wing sections.

Author Contributions: Conceptualization, J.Y.; conducted the data collection, J.Y. and Y.Y.; funding acquisition, H.Y.; supervision, N.L., H.Y. and W.Z.; formal analysis, J.Y., H.Y., W.Z.; writing—original draft preparation, J.Y.; writing—review and editing, H.Y., W.Z.; All authors have read and agreed to the published version of the manuscript.

Funding: This research was funded by Postgraduate Research & Practice Innovation Program of Jiangsu Province under grant number KYCX19_2104 and Youth Fund Project of Jiangsu Natural Science Foundation under grant number BK20170510.

Acknowledgments: The authors wish to express acknowledgement to the Postgraduate Research &Practice Innovation Program of Jiangsu Province under grant number KYCX19_2104 and Youth Fund Project of Jiangsu Natural Science Foundation under grant number BK20170510.

Conflicts of Interest: The authors declare no conflict of interest.

References

1. Liebeck, R.H. Design of subsonic airfoils for high lift. *J. Aircr.* **1978**, *15*, 547–561. [CrossRef]
2. Amini, Y.; Emdad, H.; Farid, M. Adjoint shape optimization of airfoils with attached Gurney flap. *Aerosp. Sci. Technol.* **2015**, *41*, 216–228. [CrossRef]
3. Xu, H.R.; Yang, H.; Liu, C. Numerical value analysis on aerodynamic performance of DU series airfoil with thickened trailing edge. *Trans. Chin. Soc. Agric. Eng.* **2014**, *30*, 101–108.
4. Yashodhar, V.; Humrutha, G.; Kaushik, M.; Khan, S.A. CFD Studies on Triangular Micro-Vortex Generators in Flow Control. In *IOP Conference Series: Materials Science and Engineering*; IOP Publishing: Bristol, UK, 2017; Volume 184, p. 012007.
5. Lee, T.; Su, Y.Y. Unsteady airfoil with a harmonically deflected trailing-edge flap. *J. Fluids Struct.* **2011**, *27*, 1411–1424. [CrossRef]
6. Seyednia, M.; Masdari, M.; Vakilipour, S. The influence of oscillating trailing-edge flap on the dynamic stall control of a pitching wind turbine airfoil. *J. Braz. Soc. Mech. Sci. Eng.* **2019**, *41*, 192. [CrossRef]
7. Lu, W.S.; Tian, Y.; Liu, P.P. Aerodynamic optimization and mechanism design of flexible variable camber trailing-edge flap. *Chin. J. Aeronaut.* **2017**, *30*, 988–1003. [CrossRef]
8. Zhang, W.G.; Bai, X.J.; Wang, Y.F.; Han, Y.; Hu, Y. Optimization of sizing parameters and multi-objective control of trailing edge flaps on a smart rotor. *Renew. Energy* **2018**, *129*, 75–91. [CrossRef]
9. Zhang, H.; Zhao, Z.D.; Zhou, G.X.; Kang, S. Experimental investigation of the effect of Gurney flap on DU93-W-210 airfoil aerodynamics performance. *Acta Energ. Solaris Sinica* **2017**, *38*, 601–606.
10. Amini, Y.; Liravi, M.; Izadpanah, E. The effects of Gurney flap on the aerodynamic performance of NACA 0012 airfoil in the rarefied gas flow. *Comput. Fluids* **2018**, *170*, 93–105. [CrossRef]
11. Lee, T.; Su, Y.Y. Lift enhancement and flow structure of airfoil with joint trailing-edge flap and Gurney flap. *Exp. Fluids* **2011**, *50*, 1671–1684. [CrossRef]
12. Chen, Z.J.; Stol, K.A.; Mace, B.R. Wind turbine blade optimization with individual pitch and trailing edge flap control. *Renew. Energy* **2016**, *103*, 750–765. [CrossRef]
13. Medina, A.; Ol, M.V.; Mancini, P.; Jones, A. Revisiting Conventional Flaps at High Deflection Rate. *AIAA J.* **2017**, *55*, 1–10. [CrossRef]
14. Elsayed, O.A.; Asrar, W.; Omar, A.A.; Kwon, K.; Jung, H.J. Experimental Investigation of Plain and Flapped-Wing Tip Vortex. *J. Aircr.* **2009**, *46*, 254–262. [CrossRef]
15. Little, J.; Nishihara, M.; Adamovich, I.; Saminy, M. High-lift airfoil trailing edge separation control using a single dielectric barrier discharge plasma actuator. *Exp. Fluids* **2010**, *48*, 521–537. [CrossRef]
16. Bergami, L.; Poulsen, N.K. A smart rotor configuration with linear quadratic control of adaptive trailing edge flaps for active load alleviation. *Wind Energy* **2015**, *18*, 625–641. [CrossRef]
17. Edward, T.; Christoph, B. Upstream shear-layer stabilization via self-oscillating trailing edge flaplets. *Exp. Fluids* **2018**, *59*, 145.
18. Straub, F.K.; Anand, V.R.; Lau, B.H.; Birchette, T.S. Wind Tunnel Test of the SMART Active Flap Rotor. *J. Am. Helicopter Soc.* **2018**, *63*, 012002. [CrossRef]
19. Traub, L.W. Prediction of Gurney-Flap Lift Enhancement for Airfoils and Wings. *AIAA J.* **2014**, *52*, 2087–2090. [CrossRef]
20. Lario, A.; Arina, R. Discontinuous Galerkin Method for the Study of Active Gurney Flaps. *J. Aircr.* **2017**, *54*, 1–11. [CrossRef]

21. Li, C.F.; Xu, Y.; Zhao, X.L.; Xu, J.Z. Analysis on Dynamic Performance Trailing Edge Flap on Wind Turbine Airfoil. *J. Eng. Thermophys.* **2014**, *35*, 883–887.
22. Zhu, W.J.; Behrens, T.; Shen, W.Z.; Sørensen, J.N. Hybrid Immersed Boundary Method for Airfoils with a Trailing-Edge Flap. *AIAA J.* **2013**, *51*, 30–41. [CrossRef]
23. Ng, B.F.; Palacios, R.; Kerrigan, E.C.; Graham, J.M.R.; Hesse, H. Aerodynamic load control in horizontal axis wind turbines with combined aeroelastic tailoring and trailing edge flaps. *Wind Energy* **2016**, *19*, 243–263. [CrossRef]
24. Colman, J.; Marañón Di Leo, J.; Delnero, J.S.; Martinez, M.; Boldes, U.; Bacchi, F. Lift and drag coefficients behaviour at low Reynolds number in an airfoil with Gurney flap submitted to a turbulent flow: Part 1. *Lat. Am. Appl. Res.* **2008**, *38*, 195–200.
25. Di Leo, J.M.; Martínez, M.A.; Delnero, J.S.; Saínz, M.G. Experimental study of the effect of the wake generated by oscillating Gurney flap. *Lat. Am. Appl. Res.* **2019**, *49*, 289–296.
26. Li, W.J.; Zhang, P.; Yang, S.F.; Fu, X.H.; Xiao, Y. An Experimental Method for Generating Shear-Free Turbulence Using Horizontal Oscillating Grids. *Water* **2020**, *12*, 591. [CrossRef]
27. Cheng, J.T.; Zhu, W.J.; Fischer, A.; García, N.R.; Madsen, J.; Chen, J.; Shen, W.Z. Design and validation of the high performance and low noise CQU-DTU-LN1 airfoils. *Wind Energy* **2013**, *17*, 1817–1833. [CrossRef]
28. ROHA. *LSWT Campaign Report on DTU-C21*; LM Internal Report: Jupitervej, Denmark, 2012.
29. Allen, H.J.; Vlncenti, W.G. Wall Interference in a Two-Dimensional Flow Wind Tunnel, with Consideration of the effect of Compressibility. *NACA Rep.* **1944**, *782*, 155–184.
30. Timmer, W.A.; van Rooij, R.P.J.O.M. Summary of the Delft University Wind Turbine Dedicated Airfoils. *J. Sol. Energy Eng.* **2003**, *125*, 11–21. [CrossRef]
31. Llorente, E.; Gorostidi, A.; Jacobs, M.; Timmer, W.A.; Munduate, A.; Pires, A. Wind Tunnel Tests of Wind Turbine Airfoils at High Reynolds Numbers. *J. Phys. Conf.* **2014**, *524*, 012012. [CrossRef]
32. Pires, O.; Munduate, X.; Ceyhan, O.; Jacobs, M.; Madsen, J.; Schepers, J.G. Analysis of the high Reynolds number 2D tests on a wind turbine airfoil performed at two different wind tunnels. *J. Phys. Conf. Ser.* **2016**, *749*, 012014. [CrossRef]
33. Troolin, D.R. A Quantitative Study of the Lift-Enhancing Flow Field Generated by an Airfoil with a Gurney Flap. Ph.D. Dissertation, University of Minnesota, Minnesota, MN, USA, 2009.

Publisher's Note: MDPI stays neutral with regard to jurisdictional claims in published maps and institutional affiliations.

Article

Evaluation of Tip Loss Corrections to AD/NS Simulations of Wind Turbine Aerodynamic Performance

Wei Zhong [1], Tong Guang Wang [1], Wei Jun Zhu [2,*] and Wen Zhong Shen [3]

[1] Jiangsu Key Laboratory of Hi-Tech Research for Wind Turbine Design, Nanjing University of Aeronautics and Astronautics, Nanjing 210016, China; zhongwei@nuaa.edu.cn (W.Z.); tgwang@nuaa.edu.cn (T.G.W.)
[2] College of Electrical, Energy and Power Engineering, Yangzhou University, Yangzhou 225009, China
[3] Department of Wind Energy, Technical University of Denmark, 2800 Lyngby, Denmark; wzsh@dtu.dk
* Correspondence: wjzhu@yzu.edu.cn

Received: 17 October 2019; Accepted: 12 November 2019; Published: 15 November 2019

Abstract: The Actuator Disc/Navier-Stokes (AD/NS) method has played a significant role in wind farm simulations. It is based on the assumption that the flow is azimuthally uniform in the rotor plane, and thus, requires a tip loss correction to take into account the effect of a finite number of blades. All existing tip loss corrections were originally proposed for the Blade-Element Momentum Theory (BEMT), and their implementations have to be changed when transplanted into the AD/NS method. The special focus of the present study is to investigate the performance of tip loss corrections combined in the AD/NS method. The study is conducted by using an axisymmetric AD/NS solver to simulate the flow past the experimental *NREL Phase VI* wind turbine and the virtual *NREL 5MW* wind turbine. Three different implementations of the widely used Glauert tip loss function F are discussed and evaluated. In addition, a newly developed tip loss correction is applied and compared with the above implementations. For both the small and large rotors under investigation, the three different implementations show a certain degree of difference to each other, although the relative difference in blade loads is generally no more than 4%. Their performance is roughly consistent with the standard Glauert correction employed in the BEMT, but they all tend to make the blade tip loads over-predicted. As an alternative method, the new tip loss correction shows superior performance in various flow conditions. A further investigation into the flow around and behind the rotors indicates that tip loss correction has a significant influence on the velocity development in the wake.

Keywords: wind turbine aerodynamics; actuator disc; AD/NS; tip loss correction; blade element momentum

---

1. Introduction

Wind energy is nowadays an important and increasing source of electric power. It has been the biggest contributor of renewable electricity except for hydropower, sharing about 5.5% of the global electricity production in 2018 [1]. As a primary subject of wind turbine technology, aerodynamics [2] largely determines the efficiency of wind energy extraction of an individual wind turbine or a wind farm. Along with the extensive development of high quality wind resources onshore, there are several trends in wind power industry: A large number of newly installed wind turbines have to be installed in areas with lower wind speeds and complex terrain; the layout optimization of wind turbine array becomes very important for wind farm design; offshore wind power development is accelerating, and the rotor size is continuously increasing. These trends need to be supported by more advanced and refined aerodynamic tools. More accurate aerodynamic load prediction is required for the design of a new generation of wind turbines with high efficiency and relatively low weight. Furthermore, a wind farm with dozens of wind turbines needs to be studied as a whole in order to take into account the complex interference between wind turbines.

Blade Element Momentum Theory (BEMT) [3,4] and Computational Fluid Dynamics (CFD) [5,6] are two essential methods of wind turbine aerodynamic computation. BEMT is no doubt the key method for rotor design [7], while CFD gives the most refined data of aerodynamic loads and flow parameters. A full CFD simulation with resolved rotor geometry is usually employed for an individual wind turbine [8,9]. However, full CFD is not suitable for a wind turbine array, due to the huge computational cost. The Actuator Disc/Navier-Stokes (AD/NS) method [10–13] was developed for this situation, in which the rotor geometry is not resolved, and thus, the number of mesh cells is greatly reduced. AD/NS is based on a combination of the blade-element theory and CFD. The flow field is still solved by CFD, while the rotor entity is replaced by a virtual actuator disc on which an external body force is acted. If the blade entities are represented by virtual actuator lines, it is called the Actuator Lines/Navier-Stokes (AL/NS) method [14–17]. As compared to AD/NS, the flow around the rotor solved by AL/NS is closer to the reality, since the rotor vortices are simulated. However, AL/NS requires more grid cells to describe the actuator lines, and thus, is seldom employed in wind farm simulations.

The AD/NS method has played a significant role in wind farm simulations involving wake interaction [18,19], complex terrain [20,21], atmospheric boundary layer [22,23], noise propagation [24,25], etc. Nevertheless, the AD model assumes that the number of blades is infinite, and thus, no tip loss is simulated, causing an over-prediction of the blade tip loads and power output. In order to take tip loss into account, a reliable engineering model has to be embedded into the numerical solver, which is usually called tip loss correction. However, all tip loss corrections were originally proposed for BEMT, and there is no tip loss correction specially developed for AD/NS. Most of the literature about AD/NS simulations either did not mention tip loss or declared that the Glauert tip loss correction [26] was employed. It is worth mentioning that AD/NS and BEMT solve the momentum of the flow past the rotor by using two completely different approaches. The former employs CFD, while the latter applies the momentum theory, though they share the actuator disc assumption and the blade-element theory. That leads to different implementations of tip loss correction for the two methods. In the BEMT, tip loss correction is realized by applying a correction factor F into the induced velocity through the rotor. In the AD/NS, the velocity is naturally found by the NS solver, and factor F can only be used to modify the external body force. The question is whether a tip loss correction has a good global performance when it is transplanted into AD/NS. Even for the BEMT itself, evaluations indicate that accurate tip loss correction is still not well achieved [27], and the development of new correction models is still going on [28–33]. In contrast with the massive study in the BEMT, tip loss corrections applied to AD/NS lack a comprehensive evaluation.

In the present work, the 2-Dimensional (2D) axisymmetric AD model is employed, and steady-state simulations are performed. The code solves the incompressible axisymmetric NS equations for the experimental *NREL Phase VI* wind turbine [34] and the virtual *NREL 5MW* [35] wind turbine under various axial inflow conditions. The main purpose of the numerical study is to evaluate the performance of the tip loss corrections applied to AD/NS. Three different implementations of the Glauert tip loss correction are discussed. In addition, a new tip loss correction recently proposed by Zhong et al. [33] is introduced and compared. The normal and tangential forces of blade cross-sections are chosen as the key parameters for the present evaluation. Because of the long arm of force, computational errors of the forces acted on the tip region are most likely to cause non-ignorable errors of the blade bending moment and the rotor torque (power generation). A BEMT study on the *NREL Phase VI* wind turbine by Branlard [27] showed that the power generation would be overestimated by 15% if no tip loss correction was made and a deviation of about 5% exists when various existing tip loss corrections were applied. That highlights the significance of the present study on tip loss correction.

The innovation of the present study lies in the following items. (a) Different implementations of the Glauert tip loss factor F are gathered, discussed and compared, and finally, one of them is recommended according to its best performance. Such kind of work has never been reported in the existing literature. (b) The Glauert correction is found to have a similar performance when it is transplanted from BEMT to AD/NS. (c) The new tip loss correction of Zhong et al. [33] is for the first time employed in AD/NS simulations. It is found to be generally superior to the Glauert-type corrections. That provides an alternative choice for a more accurate prediction of the blade tip loads. (d) Tip loss correction is found to have a significant influence on the velocity field, which highlights the importance of an accurate tip loss correction for not only the blade loads, but also the wake development.

The paper is organized as follows: In Section 2, the axisymmetric AD/NS method is introduced, including the governing equations of the NS approach and the formulae of the AD model; In Section 3, the Glauert and the new tip loss corrections are introduced; In Section 4, the implementations of the tip loss corrections used in AD/NS are described; In Section 5, the numerical setup, as well as the involved simulation cases, are introduced; In Section 6, all simulation results are presented and discussed; Final conclusions are made in the last section.

2. Axisymmetric AD/NS Method

2.1. Governing Equations

The governing equations of the present AD/NS simulation are the incompressible axisymmetric NS equations. In a cylindrical coordinate system as shown in Figure 1, the axial direction is defined along the z-axis, the radial direction is represented by r, and the tangential direction is represented by θ, the continuity equation is written as

$$\nabla \cdot \vec{u} = \frac{\partial u_z}{\partial z} + \frac{\partial u_r}{\partial r} + \frac{u_r}{r} = 0, \tag{1}$$

the axial and radial momentum equations read

$$\frac{\partial u_z}{\partial t} + \frac{1}{r}\frac{\partial}{\partial z}(ru_z^2) + \frac{1}{r}\frac{\partial}{\partial r}(ru_r u_z) = -\frac{1}{\rho}\frac{\partial p}{\partial z} + \frac{1}{r}\frac{\partial}{\partial z}\left[rv\left(2\frac{\partial u_z}{\partial z} - \frac{2}{3}(\nabla \cdot \vec{u})\right)\right] + \frac{1}{r}\frac{\partial}{\partial r}\left[rv\left(\frac{\partial u_z}{\partial r} + \frac{\partial u_r}{\partial z}\right)\right] + f_z, \tag{2}$$

$$\begin{aligned}\frac{\partial u_r}{\partial t} + \frac{1}{r}\frac{\partial}{\partial z}(ru_z u_r) + \frac{1}{r}\frac{\partial}{\partial r}(ru_r^2) = \\ -\frac{1}{\rho}\frac{\partial p}{\partial r} + \frac{1}{r}\frac{\partial}{\partial z}\left[rv\left(\frac{\partial u_r}{\partial z} + \frac{\partial u_z}{\partial r}\right)\right] + \frac{1}{r}\frac{\partial}{\partial r}\left[rv\left(2\frac{\partial u_r}{\partial r} - \frac{2}{3}(\nabla \cdot \vec{u})\right)\right] - 2v\frac{u_r}{r^2} + \frac{2}{3}\frac{v}{r}(\nabla \cdot \vec{u}) + \frac{u_\theta^2}{r} + f_r\end{aligned} \tag{3}$$

and the tangential momentum equation is

$$\frac{\partial u_\theta}{\partial t} + \frac{1}{r}\frac{\partial}{\partial z}(ru_z u_\theta) + \frac{1}{r}\frac{\partial}{\partial r}(ru_r u_\theta) = \frac{1}{r}\frac{\partial}{\partial z}\left(rv\frac{\partial u_\theta}{\partial z}\right) + \frac{1}{r^2}\frac{\partial}{\partial r}\left[r^3 v\frac{\partial}{\partial r}\left(\frac{u_\theta}{r}\right)\right] - \frac{u_r u_\theta}{r} + f_\theta. \tag{4}$$

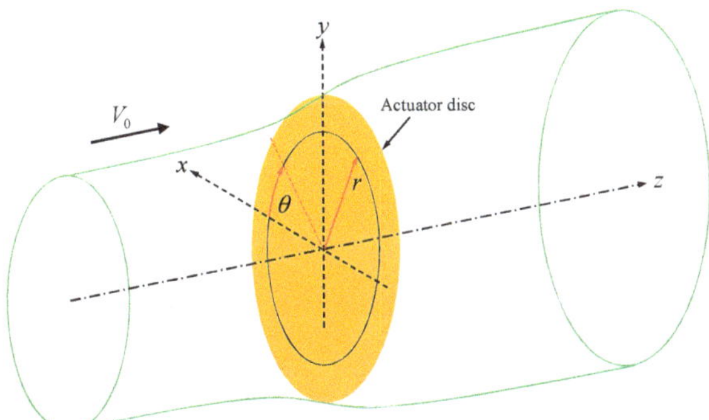

Figure 1. Definition of coordinates and the stream tube through an actuator disc.

In the above equations, $u_z/u_r/u_\theta$ is the axial/radial/tangential velocity, p is the static pressure, ρ is the air density, ν is the kinematic viscosity coefficient of air, f_z, f_r, f_θ are the source term of the axial, radial, tangential external body forces, respectively.

2.2. Force on Actuator Disc

The conceptual idea of the AD/NS method is to solve the aerodynamic force of rotor blades by using the blade-element theory and then applying its counterforce as an external body force into the momentum equations.

At each time step of solving the NS equations, the velocity passing through the actuator disc is detected and used to determine the axial induced velocity W_a and the tangential induced velocity W_t,

$$W_a = V_0 - u_z, \tag{5}$$

$$W_t = -u_\theta. \tag{6}$$

The axial interference factor a and tangential interference factor a' are then determined by,

$$a = \frac{W_a}{V_0}, \tag{7}$$

$$a' = \frac{W_t}{\Omega r}, \tag{8}$$

where V_0 is the wind speed, and Ω is the rotating speed of the wind turbine rotor.

The above interference factors are azimuthally unchanged according to the axisymmetric condition, which implies an infinite number of blades and zero tip loss. In order to estimate the tip loss of the realistic rotor with a finite number of blades, the interference factors need to be corrected as

$$\tilde{a} = f_{corr}(a), \tag{9}$$

$$\tilde{a}' = f'_{corr}(a'), \tag{10}$$

where $\tilde{a}$ and $\tilde{a}'$ denote the corrected axial and tangential interference factors, f_{corr} and f'_{corr} represent correction functions.

The axial and tangential induced velocities after the correction are written as

$$\tilde{W}_a = V_0 \tilde{a}, \tag{11}$$

$$\tilde{W}_t = \Omega r \tilde{a}'. \tag{12}$$

According to the velocity triangle shown in Figure 2, the flow angle ϕ, angle of attack α, and relative velocity V_{rel} are then determined by

$$\phi = \tan^{-1}\left(\frac{V_0 - \tilde{W}_a}{\Omega r + \tilde{W}_t}\right), \tag{13}$$

$$\alpha = \phi - \beta, \tag{14}$$

$$V_{rel}^2 = \left(V_0 - \tilde{W}_a\right)^2 + \left(\Omega r + \tilde{W}_t\right)^2, \tag{15}$$

where β is the local pitch angle of a blade cross-section.

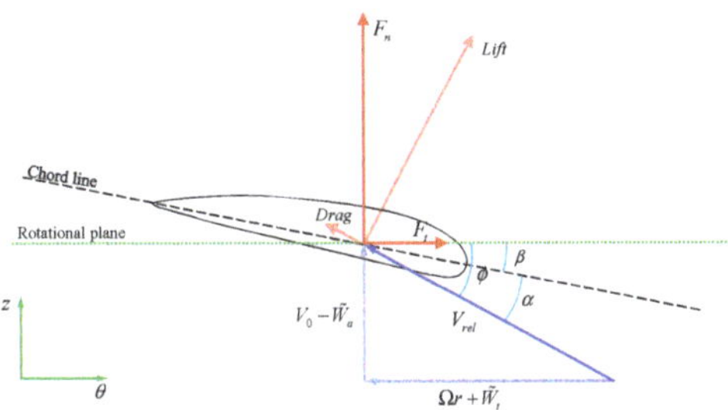

Figure 2. Illustration of the velocity and the aerodynamic force for a blade cross-section.

With the determined angle of attack and relative velocity, the lift coefficient C_l and drag coefficient C_d of each cross-section of the blade can be obtained from tabulated airfoil data. According to the relationship of force projection, the normal and tangential force coefficients are determined by

$$C_n = C_l \cos \phi + C_d \sin \phi, \tag{16}$$

$$C_t = C_l \sin \phi - C_d \cos \phi, \tag{17}$$

The axial force F_z and tangential force F_θ per radial length of the rotor are then given by

$$F_z = \frac{1}{2}\rho V_{rel}^2 cBC_n = \frac{1}{2}\rho V_{rel}^2 cB(C_l \cos \phi + C_d \sin \phi), \tag{18}$$

$$F_\theta = \frac{1}{2}\rho V_{rel}^2 cBC_t = \frac{1}{2}\rho V_{rel}^2 cB(C_l \sin \phi - C_d \cos \phi), \tag{19}$$

where c is the local chord length, and B is the number of blades. Ignoring the effect of the radial flow, the vector form of the counterforce acting on the air is

$$\vec{F} = (-F_z, 0, -F_\theta). \tag{20}$$

Rather than distributing the force only on the disc, the above force is regularized by the following Gaussian distribution along the axial direction in order to avoid a numerical singularity. The Gaussian distribution has been widely used and proven to be proper for AD/NS and AL/NS simulations [10,14–17].

$$\vec{F}_\varepsilon = \eta_\varepsilon \vec{F}, \; \eta_\varepsilon = \frac{1}{\varepsilon \sqrt{\pi}} \exp\left[-\left(\frac{z - z_0}{\varepsilon}\right)^2\right], \tag{21}$$

where z_0 is the axial position of the disc. The parameter ε serves to adjust the concentration of the regularized force, which is in the present study set to $\varepsilon = 0.02R$ where R is the rotor radius.

The source terms of the external body force in the momentum Equations (2)–(4) need to be replaced with the above force per unit volume. In the axisymmetric coordinate system, a 2D grid cell with an axial length of Δz and a radial height of Δr represents a volume of $2\pi r \Delta r \Delta z$, and thus, the force per unit volume is

$$\vec{f} = \frac{\vec{F}_\varepsilon \Delta r \Delta z}{2\pi r \Delta r \Delta z} = \frac{\vec{F}_\varepsilon}{2\pi r}, \tag{22}$$

Using Equations (20) and (21), it reads

$$\vec{f} = \frac{\eta_\varepsilon \vec{F}}{2\pi r} = \left(-\frac{\eta_\varepsilon F_z}{2\pi r}, 0, -\frac{\eta_\varepsilon F_\theta}{2\pi r}\right), \tag{23}$$

i.e.,

$$\begin{cases} f_z = -\frac{\eta_\varepsilon F_z}{2\pi r} \\ f_r = 0 \\ f_\theta = -\frac{\eta_\varepsilon F_\theta}{2\pi r} \end{cases}.$$
(24)

3. Tip Loss Corrections

3.1. Glauert Correction

The tip loss correction of Glauert [26] was developed from the study of Prandtl [36]. A function F, which was later recognized as the first tip loss factor, was derived by Prandtl making Betz's optimal circulation [37] go to zero at the blade tip,

$$F = \frac{2}{\pi}\arccos\left\{\exp\left[-\frac{B}{2}\left(1-\frac{r}{R}\right)\sqrt{1+\lambda^2}\right]\right\},$$
(25)

where B is the number of blades, r is the local radial location, R is the rotor radius, and λ is the tip speed ratio. The original derivation of Prandtl was written very briefly and was elaborated more in [4,27,38] as for a more detailed derivation.

Later on, Glauert made further contributions. First, he interpreted the physical meaning of factor F as the ratio between the azimuthally averaged induced velocity and the induced velocity at the blade position, leading to

$$F = \frac{\bar{a}}{a_B} = \frac{\bar{a'}}{a'_B},$$
(26)

where $\bar{a} = \frac{1}{2\pi}\int_0^{2\pi} a\,d\theta$ and $\bar{a'} = \frac{1}{2\pi}\int_0^{2\pi} a'\,d\theta$ are the azimuthally averaged axial and tangential interference factors, and a_B and a'_B are the interference factors at the blade position. Second, the local inflow angle ϕ was introduced into the function in order to make it consistent with the local treatment of the BEMT, leading to the following new formula of factor F,

$$F = \frac{2}{\pi}\arccos\left\{\exp\left[-\frac{B(R-r)}{2r\sin\phi}\right]\right\}.$$
(27)

The factor F was then applied into the momentum theory through a straightforward way of multiplying the axial velocity at the rotor plane with factor F, resulting in the following thrust and torque for an annular element,

$$dT = 4\pi r\rho V_0^2 a(1-a)F dr,$$
(28)

$$dM = 4\pi r^3 \rho V_0 \Omega a'(1-a)F dr.$$
(29)

Using another two equations of dT and dM derived from the blade-element theory,

$$dT = \frac{1}{2}\rho V_{rel}^2 c C_n B dr = \frac{1}{2}\rho V_{rel}^2 c(C_l\cos\phi + C_d\sin\phi)B dr,$$
(30)

$$dM = \frac{1}{2}\rho V_{rel}^2 c C_t B r dr = \frac{1}{2}\rho V_{rel}^2 c(C_l\sin\phi - C_d\cos\phi)B r dr,$$
(31)

and the velocity triangle at the rotor plane, the equations for the interference factors in the BEMT approach was derived to be:

$$\frac{a}{1-a} = \frac{\sigma(C_l\cos\phi + C_d\sin\phi)}{4F\sin^2\phi},$$
(32)

$$\frac{a'}{1+a'} = \frac{\sigma(C_l\sin\phi - C_d\cos\phi)}{4F\sin\phi\cos\phi}.$$
(33)

The above two equations lead to the final iterative formulae of the BEMT approach with the Glauert tip loss correction. Their difference from the baseline BEMT approach is only the appearance of factor F in the equations. Obviously, the application of the Glauert correction in BEMT is very simple, which is also a great advantage. There are several variations of the Glauert correction in which the way of applying factor F is different [39,40], but the original Glauert tip loss correction is the most commonly used form till today.

3.2. A Newly Developed Correction

A new tip loss correction was recently proposed for BEMT by Zhong et al. [33], based on a novel insight into tip loss. In contrast with the Prandtl/Glauert series corrections that begin with an actuator disc and estimate the effect of the finite number of blades, the new correction begins with a non-rotating blade and estimates the effect of rotation on tip loss. It has been validated in BEMT computations and showed superior performances in the cases involving flow separation or high axial interference factor.

The correction was realized by using two factors of F_R and F_S that treat the rotational effect and the 3D effect, respectively.

$$F_R = 2 - \frac{2}{\pi} \arccos\left\{ \exp\left[-2B(1 - r/R)\sqrt{1 + \lambda^2} \right] \right\},$$ (34)

$$F_S = \frac{2}{\pi} \arccos\left\{ \exp\left[-\left(\frac{1 - r/R}{\bar{c}/R} \right)^{3/4} \right] \right\}, \quad \bar{c} = \frac{S_t}{R - r},$$ (35)

where $\bar{c}$ denotes a geometric mean chord length, and S_t is the projected area of the blade between the present cross-section and the tip. The purpose of introducing $\bar{c}$ is to deal with tapered blades and those with sharp tips.

The factor F_R was applied to the BEMT approach by using:

$$\frac{aF_R(1 - aF_R)}{(1 - a)} = \frac{\sigma\left(\tilde{C}_l \cos\phi + \tilde{C}_d \sin\phi\right)}{4\sin^2\phi},$$ (36)

$$\frac{a'F_R(1 - aF_R)}{(1 + a')(1 - a)} = \frac{\sigma\left(\tilde{C}_l \sin\phi - \tilde{C}_d \cos\phi\right)}{4\sin\phi\cos\phi},$$ (37)

in which the employed lift and drag coefficients were corrected by factor F_S, rather than the direct use of the airfoil data.

$$\tilde{C}_l = \frac{1}{2}\left[C_l(\alpha)F_S + C_l^* \right],$$ (38)

$$\tilde{C}_d = \frac{1}{\cos^2\alpha_i}\left[C_d(\alpha_e)\cos\alpha_i + \left(\tilde{C}_l \cos\alpha_i + C_d(\alpha_e)\tan\alpha_i \right)\sin\alpha_i \right];$$ (39)

where

$$C_l^* = \frac{1}{\cos^2\alpha_i}\left[C_l(\alpha_e)\cos\alpha_i - C_d(\alpha_e)\sin\alpha_i \right],$$ (40)

$$\alpha_i = \frac{C_l(\alpha)}{m}(1 - F_S),$$ (41)

where m is the slope of the lift-curve of the airfoil before flow separation, α_i is called the downwash angle, and α_e is called the effective angle of attack.

By comparing with the Glauert tip loss correction, the new tip loss correction appears more complicated in application. It is simplified in the present study: First, Equations (36) and (37) will not be employed naturally in the AD/NS method; Second, Equations (38)–(41) are further simplified to

$$\tilde{C}_l = \frac{1}{2}\left[C_l(\alpha)F_S + C_l(\alpha_e) \right], \quad \alpha_e = \alpha - \alpha_i,$$ (42)

$$\tilde{C}_d = C_d(\alpha_e) + \tilde{C}_l \tan \alpha_i, \ \alpha_i = \frac{C_l(\alpha)}{m}(1 - F_S). \tag{43}$$

The simplification is derived by using the fact that α_i and C_d are relatively small. More detailed applications are shown in the next section.

4. Applying Corrections to AD/NS Simulation

4.1. Application of Glauert Tip Loss Factor F

Equations (32) and (33) where the Glauert tip loss factor F is applied are not employed in the AD/NS method because the flow is simulated by the NS solver instead of the momentum theory. As a result, an alternative way has to be employed correctly to apply the tip loss factor F. Nevertheless, among the literature studies, little literature mentions the detail of how the tip loss factor is applied to the AD/NS simulations. After an extensive literature review, we have found three representative documents in which Sørensen et al. [41], Mikkelsen [42], and Shen et al. [43] described their implementations. In order to facilitate the distinction, the implementations adopted by the three studies are denoted as Glauert-A, Glauert-B and Glauert-C in the present paper, respectively.

4.1.1. Glauert-A Correction

Sørensen et al. [41] performed a tip loss correction by applying factor F to modify the aerodynamic force that determines the external body force in the NS equations. They replaced the lift coefficient C_l by C_l/F (there was no need to modify the drag in their study as the drag was assumed not to produce the external body force). In the present Glauert-A correction, C_n and C_t, in Equations (18) and (19), are replaced by:

$$\tilde{C}_n = C_n/F, \tag{44}$$

$$\tilde{C}_t = C_t/F. \tag{45}$$

That is equivalent to replacing C_l by C_l/F and C_d by C_d/F, according to Equations (16) and (17).

Sørensen et al. [41] did not explain the reason why factor F could be directly used to modify the force. They might be inspired by Equations (32) and (33) in which the existence of F can be looked as corrections to C_n and C_t, although from a physical point of view it is a correction to the interference factors. It is important to note that the corrected force is only used for determining the external body force, so as the flowfield, while the force acting on the blade should be regarded as the original one. In addition, the interference factors in Equations (9) and (10) should no longer be corrected, leading to $\tilde{a} = a$ and $\tilde{a}' = a'$.

This implementation involves a division operation with denominator F, which causes unreasonable big values of $\tilde{C}_n$ and $\tilde{C}_t$ at the extreme tip where $F \to 0$. However, there is no exact criterion for defining what a big value is unreasonable because the $\tilde{C}_n$ and $\tilde{C}_t$ themselves are introduced as an engineering correction rather than a physical concept. Sørensen et al. [41] did not mention this problem, and no obvious numerical fluctuation is observed in his result. That is possible because the problem is limited to a very small area at the tip, and thus, the integral effect of the resulted unreasonable body force is also small.

4.1.2. Glauert-B Correction

Mikkelsen [42] compared Equations (32) and (33) to the corresponding baseline equations without factor F. The baseline equations are

$$\frac{a}{1-a} = \frac{\sigma(C_l \cos \phi + C_d \sin \phi)}{4 \sin^2 \phi}, \tag{46}$$

$$\frac{a'}{1+a'} = \frac{\sigma(C_l \sin \phi - C_d \cos \phi)}{4 \sin \phi \cos \phi}. \tag{47}$$

In order to distinguish from the parameters in the baseline equations, we here rewrite Equations (32) and (33) to

$$\frac{F\tilde{a}}{(1-\tilde{a})} = \frac{\sigma\left(\tilde{C}_l \cos\tilde{\phi} + \tilde{C}_d \sin\tilde{\phi}\right)}{4\sin^2\tilde{\phi}}, \tag{48}$$

$$\frac{F\tilde{a}'}{(1+\tilde{a}')} = \frac{\sigma\left(\tilde{C}_l \sin\tilde{\phi} - \tilde{C}_d \cos\tilde{\phi}\right)}{4\sin\tilde{\phi}\cos\tilde{\phi}}. \tag{49}$$

The implementation of Mikkelsen [42] adopts the following equations which implies an assumption of $\phi = \tilde{\phi}$ which in fact does not accurately hold,

$$\frac{a}{1-a} = \frac{\sigma(C_l \cos\phi + C_d \sin\phi)}{4\sin^2\phi} = \frac{F\tilde{a}}{(1-\tilde{a})}, \tag{50}$$

$$\frac{a'}{1+a'} = \frac{\sigma(C_l \sin\phi - C_d \cos\phi)}{4\sin\phi\cos\phi} = \frac{F\tilde{a}'}{(1+\tilde{a}')}. \tag{51}$$

The corrected interference factors were, thus, determined by

$$\tilde{a} = \frac{a}{F(1-a)+a}, \tag{52}$$

$$\tilde{a}' = \frac{a'}{F(1+a')-a'}. \tag{53}$$

The tip loss correction was completed as the above functions for $\tilde{a}$ and $\tilde{a}'$ were used to replace Equations (9) and (10).

4.1.3. Glauert-C Correction

Shen et al. [43] performed their correction by directly using Equation (26) which represents the physical meaning of factor F. Using the azimuthally uniform condition ($\bar{a} = a$), the correction was completed by replacing Equations (9) and (10) with

$$\tilde{a} = a_B = a/F, \tag{54}$$

$$\tilde{a}' = a'_B = a'/F. \tag{55}$$

Similar to the Glauert-A correction, this implementation involves a division operation with denominator F. That causes an unphysical big value of $\tilde{a}$ at the extreme tip where $F \to 0$. A limiter is set in the present study to force the result to be $\tilde{a} = 1$ when it is greater than 1. The value of $\tilde{a}'$ is usually much less than $\tilde{a}$ and is not limited here.

4.2. Application of New Correction

The application of the new correction consists of two steps: The first is using factor F_R determined by Equation (34) and the second is using factor F_S determined by Equation (35).

In the first step, an implementation similar to the Glauert-C correction is adopted, which is performed by replacing Equations (9) and (10) with

$$\tilde{a} = a/F_R, \tag{56}$$

$$\tilde{a}' = a'/F_R. \tag{57}$$

A limiter of $\tilde{a} = 1$ is used when a is greater than 1. The angle of attack α can then be determined by calculations using Equations (11)–(14).

The second step is to correct the lift and drag coefficients by using Equations (42) and (43), the result of which is used to replace the C_l and C_d in Equations (18) and (19).

5. Computational Setup

5.1. Flow Solver and Mesh Configuration

The 2D axisymmetric NS equations are solved by using an in-house code EllipSys2D [44,45] developed at Technical University of Denmark (DTU). The code is a general incompressible flow solver with multi-block and multi-grid strategy. The equations are discretized with a second-order finite volume method. In the spatial discretization, a central difference scheme is applied to the diffusive terms and the QUICK (Quadratic Upstream Interpolation for Convective Kinematics) upwind scheme is applied to the convective terms. The SIMPLEC (Semi-Implicit Method for Pressure-Linked Equations) scheme is used for the velocity-pressure decoupling. The turbulence flow is simulated using the method of Reynolds-averaged Navier-Stokes equations (RANS) in which the k-ω turbulence model of Menter [46] is employed with a modification for a better simulation of the turbulence quantities in the free-stream flow [47,48].

The coordinate for the AD/NS simulations is defined in the z-y plane. A computational mesh is generated, as shown in Figure 3 where the inflow, outflow and axisymmetric boundaries are indicated, the length of the computational domain is 30, which is nondimensionalized with the rotor radius. The blade is positioned at $z = 0$ and the grid cells are clustered around $z = 0$ and $y = 1$ to ensure a better resolution near the blade tip. The mesh is composed of four blocks where 64×64 grid points are used for each block. There are 64 grid points on the AD along the radial direction, and 20 points in the range of [-0.02, 0.02] in the axial direction where the Gaussian distribution of the body force plays a significant role. Since the axisymmetric boundary condition is applied, the 2D flow solution is regarded as an azimuthal slice of a full 3D field.

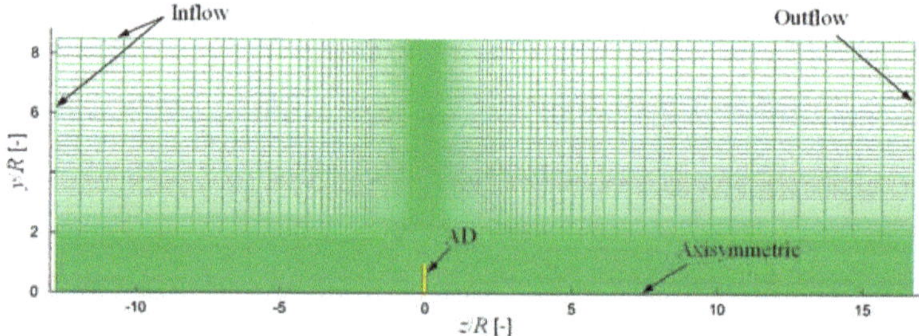

Figure 3. Mesh configuration for the Actuator Disc/Navier-Stokes (AD/NS) simulations.

The above computational setup has been proven to perform well in AD/NS simulations, as shown in the previous work of Cao et al. [49] where both blade loads and wake flows were validated against experiments.

5.2. Simulation Cases

Simulations are performed for two different wind turbines of the *NREL Phase VI* [34] and the *NREL 5MW* [35] that represent a small and a large rotor size, respectively. Additionally, the *NREL Phase VI* rotor has a very blunt blade tip shape as compared with the *NREL 5MW* rotor.

The operational conditions of the two rotors are listed in Table 1. Cases 1, 2 and 3 are set to be consistent with the *NREL UAE* experiment in axial inflow conditions [34]. Various wind speeds of 7 m/s, 10 m/s and 13 m/s are considered, while the rotating speed remains unchanged. The flow is fully attached on the blade surface at a wind speed of 7 m/s, but is partly separated at 10 m/s and 13 m/s (Higher wind speed leads to heavier flow separation) [8]. Measured pressure distributions (from which the force can be obtained by pressure integral) at five blade sections ($r/R = 0.30, 0.47, 0.63, 0.80, 0.95$) are available for these cases [50]. Cases 4 and 5 are two typical points on the designed power curve of the *NREL 5MW* wind turbine [35]. The wind speed and rotating speed of Case 5 are the rated parameters of this wind turbine. Case 4 represents a condition with a lower wind speed of 8 m/s at which the rotating speed is reduced to 9.22 rpm for tracking the optimum tip speed ratio.

Table 1. Simulation cases for two different wind turbines.

Rotor Name	Number	Wind Speed (V_0, m/s)	Rotating Speed (Ω, rpm)	Tip Speed Ratio (λ)	Tip Pitch Angle (°)
NREL Phase VI	1	7.0	72	5.4	3
	2	10.0	72	3.8	3
	3	13.0	72	2.9	3
NREL 5MW	4	8.0	9.22	7.60	0
	5	11.4	12.06	6.98	0

These cases cover the following multiple situations: Wind turbines with remarkably different rotor sizes and tip shapes; flow with fully attached and separated conditions; operating conditions with various wind speeds and rotating speeds. It is, therefore, interesting to see the performance of the tip loss corrections in these cases.

Experimental data are preferred as the reference for the computational results of the *NREL Phase VI* rotor, as well as the full CFD data are also displayed in some results. For the cases of the *NREL 5MW* rotor, there is no experimental data available such that the full CFD data are the only reference. The related full CFD simulations were previously conducted by the authors, see Zhong et al. [8,33].

6. Simulation Results

6.1. Results of Glauert-Type Corrections

6.1.1. Glauert-A/B/C Corrections

The blade loads for Cases 1, 2 and 3 are gathered in Figure 4, which represent results obtained from the *NREL Phase VI* rotor. The *NREL Phase VI* blade has a linear change of chord distribution starting from $r = 1.3$ m with a constant slope of 0.1. The chord length is 0.358 m at the tip, which is about 48% of the largest chord length in the blade inboard part. Therefore, the loads do not converge to zero without a tip loss correction.

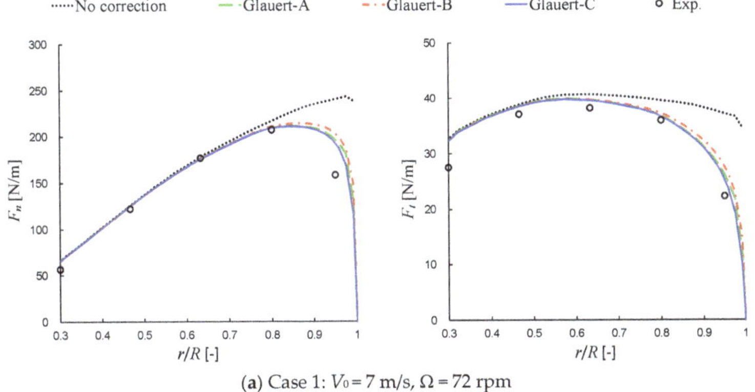

(a) Case 1: $V_0 = 7$ m/s, $\Omega = 72$ rpm

Figure 4. *Cont.*

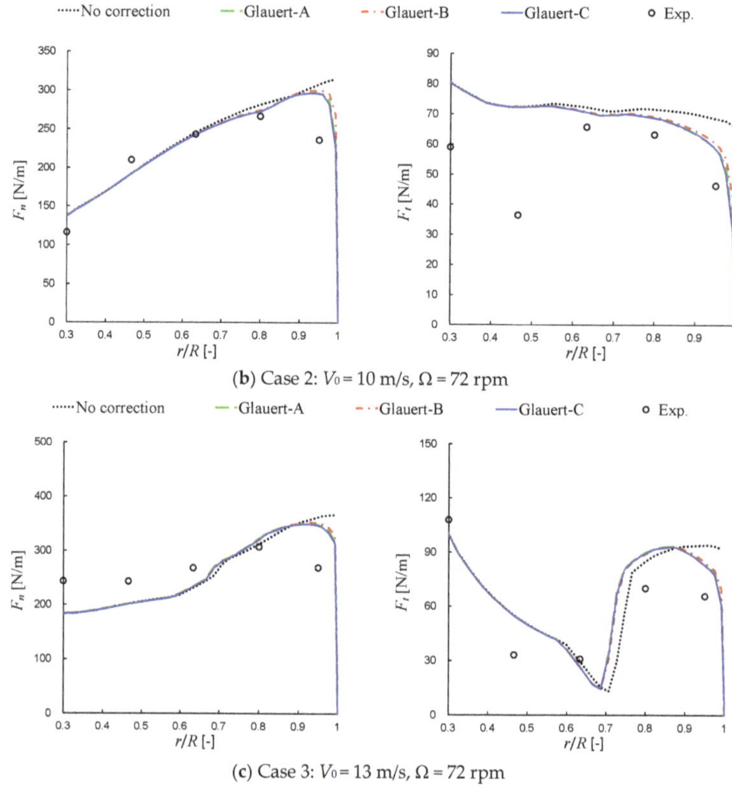

Figure 4. Normal force F_n and tangential force F_t along the *NREL Phase VI* blade.

It is noticed that there is a certain degree of difference between the results of the Glauert-A/B/C corrections in the tip region (e.g., at $r/R > 0.8$). In general, Glauert-C leads to lower loads which are closer to the experimental data, Glauert-A gives results close to, but slightly higher than, those of Glauert-C, while Glauert-B results in relatively higher loads near the tip. Taking the F_n in Figure 4a as an example of quantitative comparison, the relative difference between the results of Glauert-B and Glauert-C is about 4% at $r/R = 0.95$. The difference in other cases is not larger than this value.

At a wind speed of 7 m/s, all the curves of F_n generally agree with the five experimental data except that at $r/R = 0.95$ where an overestimation is observed. This overestimation becomes much more remarkable as the wind speed increases to 10 m/s and 13 m/s. At 10 m/s, there is no much difference around $r/R = 0.9$ between the results with no correction and with the Glauert-type corrections, indicating that almost no effective correction is made here. At 13 m/s, the curves of the Glauert-type corrections even exceed the uncorrected curve in a r/R range of about [0.66, 0.88]. It is clear that the corrections perform much worse at the higher wind speeds as compared with 7 m/s. The overestimation at $r/R = 0.95$ is also observed in the results for F_t, although it looks better to some extent, especially at 7 m/s.

Considering the fact that the Glauert correction was derived based on the potential hypothesis, it is reasonable to believe that the unusually poor performance at the higher wind speed is caused by the flow separation. According to the results of the *NREL UAE Phase VI* experiments [50], the angle of attack at most cross-sections of the blade exceeds the linear range of their aerodynamic polar when the wind speed is increased to be higher than 10m/s. Taking the cross-section of $r/R = 0.63$ as an example, the angles of attack are 5.9°, 11.8°, 16.9° at a wind speed of 7 m/s, 10 m/s, 13 m/s, respectively. The exact operating points for three cases on the lift polar of the S809 airfoil (the airfoil of all cross-sections of the *NREL Phase VI* blade) are depicted in Figure 5. It clearly shows that Case 2 and Case 3 are in the nonlinear lift region corresponding to flow separation, while Case 1 is in the linear lift region corresponding to attached flow, which implies that flow separation occurs at 10 m/s and 13 m/s.

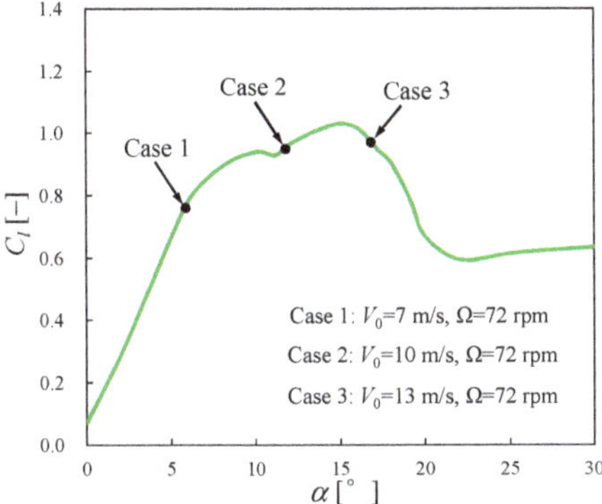

Figure 5. Operating points of the cross-section of r/R = 0.63 on the lift polar of the S809 airfoil. (The Reynolds number of the airfoil is 1×10^6 which is close to that of the cross-section).

The normal and tangential forces along a blade of the *NREL 5MW* rotor are plotted in Figure 6. As a widely studied virtual rotor, full CFD solutions [33] are often used as the reference data. Although the typical rotating speed of the *NREL 5MW* rotor is a few times less than the *NREL Phase VI* rotor, the resulted tip speed ratio is higher, which is closer to the optimal design point. The blade tip of the *NREL 5MW* rotor is smoothly sharpened where the slope at the last 4 m is 0.46, and the chord length at the tip is only 4.3% of the largest chord length in the blade inboard part. It can be seen in Figure 6 that the loads at the tip naturally approach to zero even without a tip loss correction.

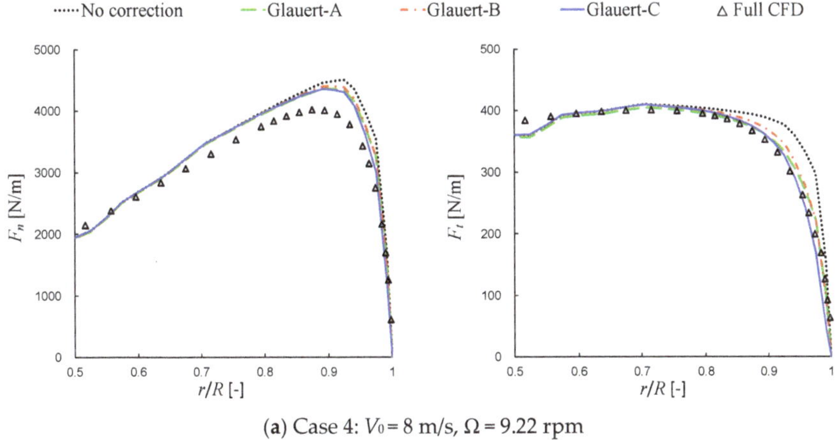

(**a**) Case 4: V_0 = 8 m/s, Ω = 9.22 rpm

Figure 6. *Cont.*

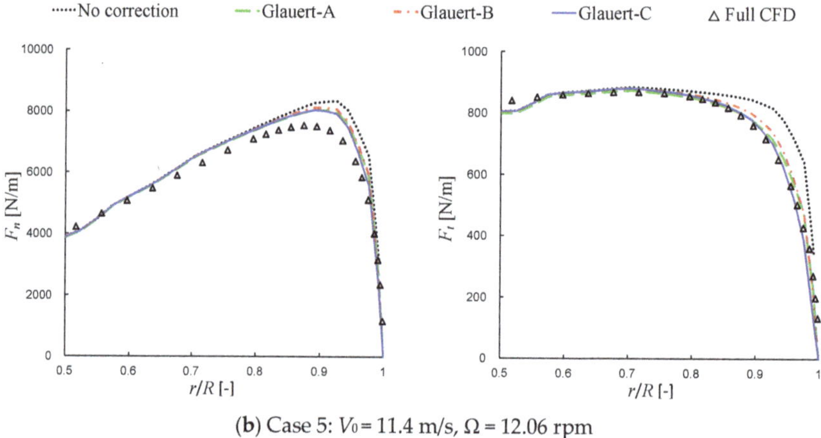

(**b**) Case 5: V_0 = 11.4 m/s, Ω = 12.06 rpm

Figure 6. Normal force F_n and tangential force F_t along the *NREL 5MW* blade.

As a large wind turbine with pitch control, there is no flow separation on the most area of the blade, including the tip. That means the flow pattern ideally keeps the same on the blade as the wind speed increases, which is significantly different from the situation of the *NREL Phase VI* rotor. As seen in Figure 6, by increasing the wind speed from 8 m/s to 11.4 m/s, the forces are greatly increased, whereas, their distributions near the tip region remain similar. All the Glauert-A/B/C corrections over-predict the normal forces near the blade tip, whereas, better agreement with the reference data is found for the tangential forces. Such a result is to some extent similar to that of the *NREL Phase VI* rotor at 7 m/s. Slight discrepancies are also noticed between the three Glauert-type corrections, and the Glauert-C performs better among them.

6.1.2. Comparison with BEMT Results

The Glauert tip loss correction was originally proposed for BEMT. As applied to AD/NS simulations, the correction factor F works in a different way. It is worth to mention that the specific difference in the final results would be caused by the transplantation from BEMT to AD/NS. In order to perform a comparison, BEMT computations under the same conditions of the present study are conducted.

The comparisons were shown by percentages, which represent the relative change of blade loads before and after using a tip loss correction,

$$\delta_n = \frac{F_{n,No} - F_{n,Gl}}{F_{n,No}} \times 100\%,$$ (58)

$$\delta_t = \frac{F_{t,No} - F_{t,Gl}}{F_{t,No}} \times 100\%,$$ (59)

where $F_{n,No}$ and $F_{n,Gl}$ are the normal forces obtained from computations with no tip loss correction and with the Glauert correction (the standard Glauert correction in BEMT and the Glauert-type corrections in AD/NS), respectively, $F_{t,No}$ and $F_{t,Gl}$ are the corresponding tangential forces.

The above parameters are calculated independently using BEMT and AD/NS. The results are shown in Figure 7 where the Glauert-BEMT denotes the results for the standard Glauert correction used in BEMT, and the Glauert-A/B/C denotes the results for the Glauert-type corrections used in AD/NS. A certain degree of difference between the Glauert-BEMT and the Glauert-A/B/C results is observed. The Glauert-C again shows the best performance, since it is generally most consistent with the Glauert-BEMT, which means the transplantation of the tip loss correction from BEMT to AD/NS does not notably influence its performance. The results for Cases 3 and 4 also agree with the above conclusion, which is not displayed here for simplicity.

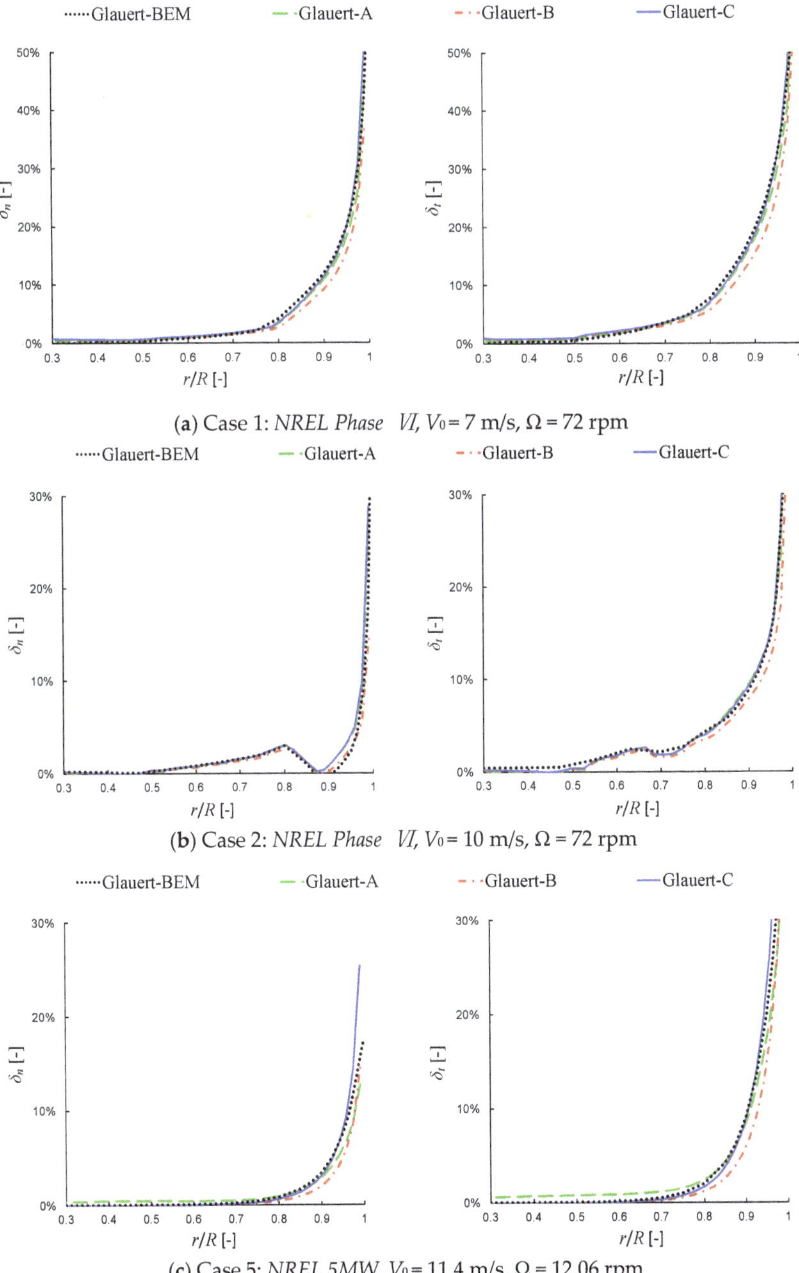

(**a**) Case 1: *NREL Phase VI*, $V_0 = 7$ m/s, $\Omega = 72$ rpm

(**b**) Case 2: *NREL Phase VI*, $V_0 = 10$ m/s, $\Omega = 72$ rpm

(**c**) Case 5: *NREL 5MW*, $V_0 = 11.4$ m/s, $\Omega = 12.06$ rpm

Figure 7. Distributions of δ_n and δ_t along a rotor blade.

6.2. Results of New Tip Loss Correction

The simplified form (see Section 3.2) of the new correction developed by Zhong et al. [33] is used in the present simulations. The results of blade loads for Cases 1, 2 and 5 are displayed in Figure 8a–c, respectively. Case 1 represents a typical condition of the *NREL Phase VI* wind turbine with the fully attached flow on its blades, while Case 2 represents another condition of flow separation (stall). Case 5 represents the rated operating condition of the *NREL 5MW* wind turbine. The results for Cases 3 and 4 are not displayed here for simplicity. (The performance of the correction in Case 3 is similar to that in Case 2, while the performance in Case 4 is similar to that in Case 5.) The Glauert-C correction, which performs best in Glauert-A/B/C, is compared with the new correction. Full CFD results [33] are employed as the reference data for the *NREL 5MW* wind turbine. As compensation for the sparse points of the experimental data, full CFD data [8,33] are also employed for the *NREL Phase VI* wind turbine.

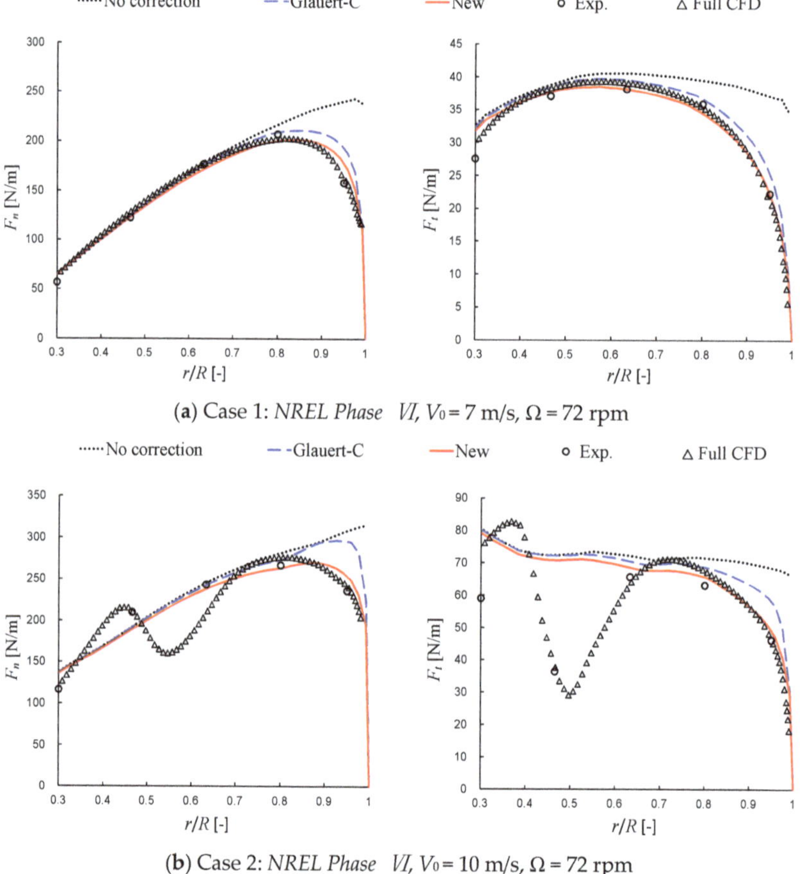

(**a**) Case 1: *NREL Phase VI*, $V_0 = 7$ m/s, $\Omega = 72$ rpm

(**b**) Case 2: *NREL Phase VI*, $V_0 = 10$ m/s, $\Omega = 72$ rpm

Figure 8. *Cont.*

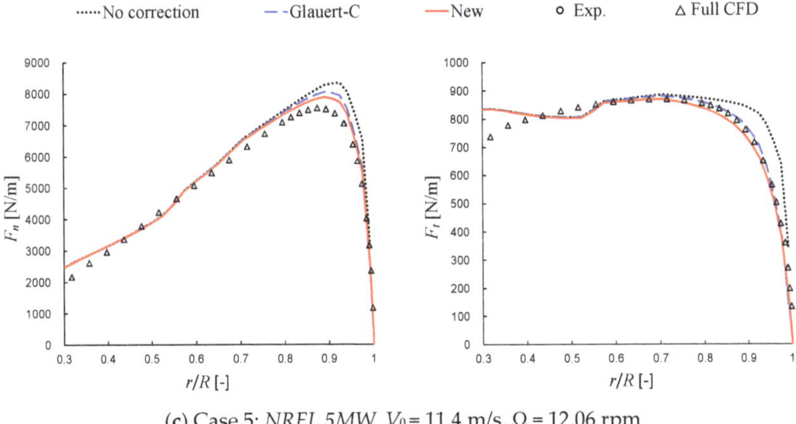

(**c**) Case 5: *NREL 5MW*, V_0 = 11.4 m/s, Ω = 12.06 rpm

Figure 8. Normal force F_n and tangential force F_t along the *NREL 5MW* blade.

It is seen in the result of Case 1, as shown in Figure 8a, that the full CFD result achieves the best agreement with the experimental data and the new correction is superior to the Glauert-C correction. As compared to the Glauert-C correction, the overestimation of F_n in the tip region is properly corrected towards the reference data by the new correction. A similar tendency is seen from the tangential force distribution. In Case 2, as shown in Figure 8b, the new correction maintains its accuracy, while the Glauert-C correction produces larger positive error, due to the occurrence of flow separation. The performance of the two corrections is similar to those in BEMT computations [33].

For the *NREL 5MW* rotor, as shown in Figure 8c, the new correction again makes a better correction to the blade loads in the tip region. The corrected curve of F_n for the new correction is closer to the reference data compared to that for the Glauert-C correction. The difference between the two corrections in tangential force (F_t) is small, and both the corrections can be considered to be doing well.

6.3. Comparison of Velocity Field

The blade loads are affected by the local flow around the blade, and the forces consequently modify the flow around the blade and further in the wake. The flowfield simulated by the AD/NS demonstrates how the tip loss corrections affect the velocity field. In order to clearly show the influence of using different tip loss corrections, instead of directly showing the velocity field, the following velocity difference is defined,

$$\delta u_z = \frac{\left|u_{z,new} - u_{z,Gl}\right|}{u_{z,new}} \times 100\%,\tag{60}$$

where $u_{z,new}$ and $u_{z,Gl}$ are the axial velocity obtained from the simulations using the new tip loss correction and the Glauert-C tip loss correction, respectively.

Figure 9 displays the contours of δu_z in the flow field of the studied two rotors, which highlights the influence of the tip loss corrections in terms of changing the velocity field not only around the blade tip, but also in the wake. The influence of δu_z is more concentrated near the blade tip in the rotor plane, whereas, the inner part is hardly affected. In the wake, the influence range of δu_z tends to cover a wider radial space downstream. The magnitude is even increased in the wake of the *NREL 5MW* rotor.

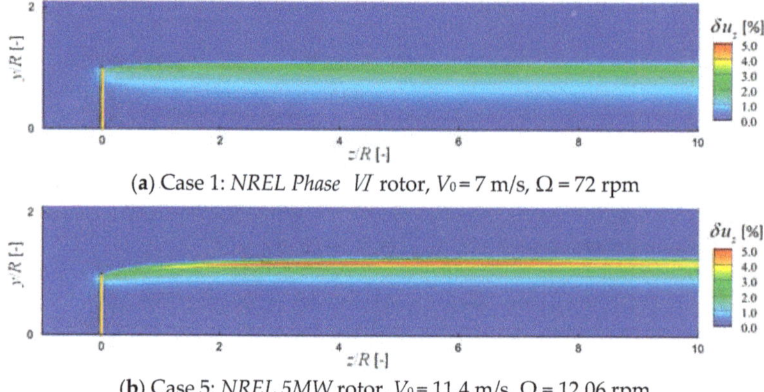

(**a**) Case 1: *NREL Phase VI* rotor, $V_0 = 7$ m/s, $\Omega = 72$ rpm

(**b**) Case 5: *NREL 5MW* rotor, $V_0 = 11.4$ m/s, $\Omega = 12.06$ rpm

Figure 9. Contours of the axial velocity difference between the results of the new and the Glauert-C tip loss corrections.

7. Conclusions

This work presented AD/NS simulations for the *NREL Phase VI* and the *NREL 5MW* rotors, which represent the typical small size and modern size rotors. The primary purpose of the numerical study is to evaluate the performance of tip loss corrections employed in the AD/NS simulations. Three different implementations of the widely used Glauert tip loss factor *F*, denoted as Glauert-A/B/C, are discussed and evaluated in the study. In addition, a newly developed tip loss correction is also evaluated. As an extension, the influence of employing different tip loss corrections to the velocity field is studied as well. The following conclusions are drawn from the present study:

- The three different implementations of the Glauert tip loss factor showed a certain degree of difference to each other, although the relative difference in blade loads is generally no more than 4%. The Glauert-C correction, in which the tip loss factor *F* is directly used to be divided by the interference factors, is recommended for AD/NS simulations, since it gives results closest to the reference data in all the studied cases.
- The performance of the Glauert-C correction in AD/NS was found to be almost equal to that of the standard Glauert correction in BEMT. That provides evidence for the reasonableness of the transplantation of the tip loss correction from BEMT into AD/NS.
- The Glauert-type corrections tend to make an over-prediction of the blade tip loads. In contrast, the new correction showed superior performance in various conditions of the present study.
- A difference in tip loss correction leads to the influence of not only on the blade tip loads, but also on the velocity field around and after the rotor. The range of this influence is observed expanding in the rotor wake. In addition, the magnitude of this influence is found to be increased after the *NREL 5 MW* rotor.

The above observations help to understand the performance of different tip loss corrections employed in AD/NS simulations. It is indicated that the blade loads, as well as the wake velocity, can be simulated better by choosing the best implementation of the Glauert tip loss factor or employing the new tip loss correction. That is the major contribution of this study.

Author Contributions: Conceptualization, W.Z.; Formal analysis, W.Z. and W.J.Z.; Funding acquisition, T.G.W.; Investigation, W.Z. and W.J.Z.; Methodology, W.Z.; Supervision, T.G.W. and W.Z.S.; Writing—Original draft, W.Z.; Writing—Review and editing, W.J.Z. and W.Z.S.

Funding: This work was funded jointly by the National Natural Science Foundation of China (No. 51506088 and No. 11672261), the National Basic Research Program of China ("973" Program) under Grant No. 2014CB046200, the Priority Academic Program Development of Jiangsu Higher Education Institutions, and the Key Laboratory of Wind Energy Utilization, Chinese Academy of Sciences (No. KLWEU- 2016-0102).

Conflicts of Interest: The authors declare no conflict of interest.

References

1. *Renewables 2019 Global Status Report*; REN21: Paris, France, 2019.
2. Shen, W.Z. Special issue on wind turbine aerodynamics. *Appl. Sci.* **2019**, *9*, 1725. [CrossRef]
3. Hansen, M.O.L. The classical blade element momentum method. In *Aerodynamics of Wind Turbines*, 3rd ed.; Routledge: Abingdon, UK, 2015; pp. 38–53.
4. Sørensen, J.N. *General Momentum Theory for Horizontal Axis Wind Turbines*; Springer International Publishing Switzerland: Cham, Switzerland, 2016; pp. 123–132. [CrossRef]
5. Sumner, J.; Watters, C.S.; Masson, C. CFD in wind energy: The virtual, multiscale wind tunnel. *Energies* **2010**, *3*, 989–1013. [CrossRef]
6. Sanderse, B.; van er Pijl, S.P.; Koren, B. Review of computational fluid dynamics for wind turbine wake aerodynamics. *Wind Energy* **2011**, *14*, 799–819. [CrossRef]
7. Rehman, S.; Alam, M.M.; Alhems, L.M.; Rafique, M.M. Horizontal axis wind turbine blade design methodologies for efficiency enhancement—A review. *Energies* **2018**, *11*, 506. [CrossRef]
8. Zhong, W.; Tang, H.W.; Wang, T.G.; Zhu, C. Accurate RANS simulation of wind turbines stall by turbulence coefficient calibration. *Appl. Sci.* **2018**, *8*, 1444. [CrossRef]
9. Rodrigues, R.V.; Lengsfeld, C. Development of a computational system to improve wind farm layout, part I: Medel validation and near wake analysis. *Energies* **2019**, *12*, 940. [CrossRef]
10. Mikkelsen, R.; Sørensen, J.N.; Shen, W.Z. Modelling and analysis of the flow field around a coned rotor. *Wind Energy* **2001**, *4*, 121–135. [CrossRef]
11. Sharpe, D.J. A general momentum theory applied to an energy-extracting actuator disc. *Wind Energy* **2004**, *7*, 177–188. [CrossRef]
12. Moens, M.; Duponcheel, M.; Winckelmans, G.; Chatelain, P. An actuator disk method with tip-loss correction based on local effective upstream velocities. *Wind Energy* **2018**, *21*, 766–782. [CrossRef]
13. Simisiroglou, N.; Polatidis, H.; Ivanell, S. Wind farm power production assessment: Introduction of a new actuator disc method and comparison with existing models in the context of a case study. *Appl. Sci.* **2019**, *9*, 431. [CrossRef]
14. Sørensen, J.N.; Shen, W.Z. Numerical modeling of wind turbine wakes. *J. Fluids Eng.* **2002**, *124*, 393–399. [CrossRef]
15. Shen, W.Z.; Zhu, W.J.; Sørensen, J.N. Actuator line/Navier-Stokes computations for the MEXICO rotor: Comparison with detailed measurements. *Wind Energy* **2012**, *15*, 811–825. [CrossRef]
16. Martínez-Tossas, L.A.; Churchfield, M.J.; Meneveau, C. A highly resolved large-eddy simulation of a wind turbine using an actuator line model with optimal body force projection. *J. Phys. Conf. Ser.* **2016**, *753*, 082014. [CrossRef]
17. Martínez-Tossas, L.A.; Meneveau, C. Filtered lifting line theory and application to the actuator line model. *J. Fluid Mech.* **2019**, *863*, 269–292. [CrossRef]
18. Castellani, F.; Vignaroli, A. An application of the actuator disc model for wind turbine wakes calculations. *Appl. Energy* **2013**, *101*, 432–440. [CrossRef]
19. Wu, Y.T.; Porte-Agel, F. Modeling turbine wakes and power losses within a wind farm using LES: An application to the Horns Rev offshore wind farm. *Renew. Energy* **2015**, *75*, 945–955. [CrossRef]
20. Sessarego, M.; Shen, W.Z.; van der Laan, M.P.; Hansen, K.S.; Zhu, W.J. CFD simulations of flows in a wind farm in complex terrain and comparisons to measurements. *Appl. Sci.* **2018**, *8*, 788. [CrossRef]
21. Diaz, G.P.N.; Saulo, A.C.; Otero, A.D. Wind farm interference and terrain interaction simulation by means of an adaptive actuator disc. *J. Wind Eng. Ind. Aerodyn.* **2019**, *186*, 58–67. [CrossRef]
22. Gargallo-Peiró, A.; Avila, M.; Owen, H.; Prieto-Godino, L.; Folcha, L. Mesh generation, sizing and convergence for onshore and offshore wind farm Atmospheric Boundary Layer flow simulation with actuator discs. *J. Comput. Phys.* **2018**, *375*, 209–227. [CrossRef]
23. Mao, X.; Sørensen, J.N. Far-wake meandering induced by atmospheric eddies in flow past a wind turbine. *J. Fluid Mech.* **2018**, *846*, 190–209. [CrossRef]

24. Zhu, W.J.; Shen, W.Z.; Barlas, E.; Bertagnolio, F.; Sørensen, J.N. Wind turbine noise generation and propagation modeling at DTU Wind Energy: A review. *Renew. Sustain. Energy Rev.* **2018**, *88*, 133–150. [CrossRef]

25. Shen, W.Z.; Zhu, W.J.; Barlas, E; Barlas, E.; Li, Y. Advanced flow and noise simulation method for wind farm assessment in complex terrain. *Renew. Energy* **2019**, *143*, 1812–1825. [CrossRef]

26. Glauert, H. Airplane propellers. In *Aerodynamic Theory*; Durand, W.F., Ed.; Dover: New York, NY, USA, 1963; pp. 169–360.

27. Branlard, E. Wind Turbine Tip-Loss Corrections. Master's Thesis, Technical University of Denmark, Lyngby, Denmark, 2011.

28. Shen, W.Z.; Mikkelsen, R.; Sørensen, J.N.; Bak, C. Tip loss corrections for wind turbine computations. *Wind Energy* **2005**, *8*, 457–475. [CrossRef]

29. Sørensen, J.N.; Dag, K.O.; Ramos-Garía, N. A new tip correction based on the decambering approach. *J. Phys. Conf. Ser.* **2015**, *524*, 012097. [CrossRef]

30. Wood, D.H.; Okulov, V.L.; Bhattacharjee, D. Direct calculation of wind turbine tip loss. *Renew. Energy* **2016**, *95*, 269–276. [CrossRef]

31. Schmitz, S.; Maniaci, D.C. Methodology to determine a tip-loss factor for highly loaded wind turbines. *AIAA J.* **2017**, *55*, 341–351. [CrossRef]

32. Wimshurst, A.; Willden, R.H.J. Computatioanl observations of the tip loss mechanism experienced by horizontal axis rotors. *Wind Energy* **2018**, *21*, 544–557. [CrossRef]

33. Zhong, W.; Shen, W.Z.; Wang, T.; Li, Y. A tip loss correction model for wind turbine aerodynamic performance prediction. *Renew. Energy* **2020**, *147*, 223–238. [CrossRef]

34. Hand, M.M.; Simms, D.A.; Fingersh, L.J.; Jager, D.W.; Cotrell, J.R.; Schreck, S.; Larwood, S.M. *Unsteady Aerodynamics Experiment Phase Vi: Wind Tunnel Test Configurations and Available Data Campaigns*; NREL/TP-500-29955; National Renewable Energy Laboratory: Golden, CO, USA, 2001.

35. Jonkman, J.; Butterfield, S.; Musial, W.; Scott, G. *Definition of A 5-Mw Reference Wind Turbine for Offshore System Development*; Technical Report NREL/TP-500-38060; National Renewable Energy Laboratory: Golden, CO, USA, 2009.

36. Prandtl, L. *Applications of Modern Hydrodynamics to Aeronautics*; NACA: Washington DC, USA, 1921; No. 116.

37. Prandtl, L.; Betz, A. *Vier Abhandlungen zur Hydrodynamik und Aerodynamik*; Göttingen Nachrichten: Göttingen, Germany, 1927. (In German)

38. Ramdin, S.F. Prandtl Tip Loss Factor Assessed. Master's Thesis, Delft University of Technology, Delft, The Netherlands, 2017.

39. Wilson, R.E.; Lissaman, P.B.S. *Applied Aerodynamics of Wind Power Machines*; NSF/RA/N-74113; Oregon State University: Corvallis, OR, USA, 1974.

40. De Vries, O. *Fluid Dynamic Aspects of Wind Energy Conversion*; AGARD: Brussels, Belgium, 1979; AG-243, Chapter 4; pp. 1–50.

41. Sørensen, J.N.; Kock, C.W. A model for unsteady rotor aerodynamics. *J. Wind Eng. Ind. Aerodyn.* **1995**, *58*, 259–275. [CrossRef]

42. Mikkelsen, R. Actuator Disc Methods Applied to Wind Turbines. Ph.D. Thesis, Technical University of Denmark, Lyngby, Denmark, 2003.

43. Shen, W.Z.; Sørensen, J.N.; Mikkelsen, R. Tip loss correction for Actuator/Navier-Stokes computations. *J. Sol. Energy Eng.* **2005**, *127*, 209–213. [CrossRef]

44. Michelsen, J.A. *Basis3d—A Platform for Development of Multiblock PDE Solvers*; Technical Report AFM; Technical University of Denmark: Lyngby, Denmark, 1992.

45. Sørensen, N.N. *General Purpose Flow Solver Applied Over Hills*; RISØ-R-827-(EN); Risø National Laboratory: Roskilde, Denmark, 1995.

46. Menter, F.R. Two-equation eddy-viscosity turbulence models for engineering applications. *AIAA* **1994**, *32*, 1598–1605. [CrossRef]

47. Tian, L.L.; Zhu, W.J.; Shen, W.Z.; Sørensen, J.N.; Zhao, N. Investigation of modified AD/RANS models for wind turbine wake predictions in large wind farm. *J. Phys. Conf. Ser.* **2014**, *524*, 012151. [CrossRef]

48. Prospathopoulos, J.M.; Politis, E.S. Evaluation of the effects of turbulence model enhancements on wind turbine wake predictions. *Wind Energy* **2011**, *14*, 285–300. [CrossRef]

49. Cao, J.; Zhu, W.; Shen, W; Sørensen, J.N.; Wang, T. Development of a CFD-based wind turbine rotor optimization tool in considering wake effects. *Appl. Sci.* **2018**, *8*, 1056. [CrossRef]
50. Jonkman, J.M. *Modeling of the UAE Wind Turbine for Refinement of FAST_AD*; Technical Report NREL/TP-500-34755; National Renewable Energy Lab.: Golden, CO, USA, 2003.

Article

Development of an Advanced Fluid-Structure-Acoustics Framework for Predicting and Controlling the Noise Emission from a Wind Turbine under Wind Shear and Yaw

Mingyue Zhou [1], Matias Sessarego [1], Hua Yang [2] and Wen Zhong Shen [1,*]

[1] Fluid Mechanics Section, Department of Wind Energy, Technical University of Denmark, Nils Koppels Allé, Building 403, 2800 Lyngby, Denmark; 1993.zhou@163.com (M.Z.); msessare@gmail.com (M.S.)
[2] School of Hydraulic Energy and Power Engineering, Yangzhou University, Yangzhou 225127, China; yanghua@yzu.edu.cn
* Correspondence: wzsh@dtu.dk; Tel.: +45-4525-4317

Received: 28 September 2020; Accepted: 26 October 2020; Published: 28 October 2020

Abstract: Noise generated from wind turbines is a big challenge for the wind energy industry to develop further onshore wind energy. The traditional way of reducing noise is to design low noise wind turbine airfoils and blades. A wind turbine operating under wind shear and in yaw produces periodic changes of blade loading, which intensifies the amplitude modulation (AM) of the generated noise, and thus can give more annoyance to the people living nearby. In this paper, the noise emission from a wind turbine under wind shear and yaw is modelled with an advanced fluid-structure-acoustics framework, and then controlled with a pitch control strategy. The numerical tool used in this study is the coupled Navier–Stokes/Actuator Line model EllipSys3D/AL, structure model FLEX5, and noise prediction model (Brooks, Pope and Marcolini: BPM) framework. All simulations and tests were made on the NM80 wind turbine equipped with three blades made by LM Wind Power. The coupled code was first validated against field load measurements under wind shear and yaw, and a fairly good agreement was obtained. The coupled code was then used to study the noise source control of the turbine under wind shear and yaw. Results show that in the case of a moderate wind shear with a shear exponent of 0.3, the pitch control strategy can reduce the mean noise emission about 0.4 dB and reduce slightly the modulation depth that mainly occurs in the low-frequency region.

Keywords: aeroacoustics; wind turbine; noise modelling; noise control

1. Introduction

As a clean and renewable energy, wind power has the advantages of low cost, low environmental pollution and wide availability. These unique advantages of wind power make it an important part of the sustainable energy mix in many countries. However, wind energy also has some drawbacks that hinder its global use. The noise caused by wind turbines has become one of the primary sources polluting the urban environment today. Wind turbine noise caused by the movement of the blades through the air is often seen as an essential aspect causing great annoyance as compared with other noise sources [1]. In Danish regulation [2], a modern wind turbine should be set up at least four times its tip height away from residential areas. Even segregated by such a long distance, a turbine still produces a sound pressure level (SPL) of more than 40 dB (A), which almost equals some common home appliances [3]. A wind turbine operating under wind shear and in yaw produces periodic changes of blade loading, which intensifies the amplitude modulation (AM) of the generated noise, and thus can give more annoyance to the people living nearby. If the aerodynamic noise of a wind turbine can be reduced without limiting its rotational speed, wind energy can be utilized with a high efficiency.

Therefore, in this paper, aerodynamic characteristics and noise features of a wind turbine in wind shear and yaw is studied.

Noise from a wind turbine or wind farm often has a feature of amplitude modulation with a varying sound pressure amplitude during its operation. There are two types of AM noise: the one is created by changing the blade noise directivity to receiver [4] and the other is caused by non-uniform inflow and operation, for example the operation in wind shear and yaw. There are three methods (Japanese F-S method [5], UK method [6] and min-max method [7]) to characterize the modulation depth. Recently, Barlas et al. [8] studied the amplitude modulation caused by a wind turbine wake using an advanced Computational Fluid Dynamics (CFD) (EllipSys3D/AL) and a sound propagation model based on solving parabolic wave equation and found that the modulation depth can reach 4–5 dB even with the same strength of noise source.

The prediction formulae of trailing edge noise were summarized first using data from an experiment by Schlinker and Amiet in 1981 [9]. In 1985, Grosveld [10] used the helicopter noise formulae synthesized by Schlinker to calculate the emission noise of several two-blade, low-power downwind horizontal axis wind turbines, and the results were in good agreement with the measured data. In 1986, De Wolf continued the research based on a similar model [11]. Back to 1981, Viterna [12] utilized an approach for the low-frequency noise of a wind turbine. A rather complex airfoil self-noise prediction model (known as the Brooks, Pope and Marcolini: BPM model) using wind tunnel acoustic measurements of National Advisory Committee for Aeronautics (NACA) 0012 airfoils was published by Brooks et al. [13] in 1989. In 1996, Fuglsang and Madsen [14] used the BPM model and the inflow noise model by Lowson [15] modified from Amiet's model [16] to calculate the noise emission of a Vestas V27 wind turbine. In 2003, the National Renewable Energy Laboratory (NREL) [17] developed its NREL AirFoil Noise (NAFNoise) based on the BPM model. The BPM was improved by including the real blade geometry for predicting wind turbine noise generation by Zhu et al. from Technical University of Denmark (DTU) in 2005 [18]. In 2009, Bowdler [19] studied wind shear effects on noise at different locations and at different time instants of the day and of the year using field data collected at wind farm sites. Oerlemans [20] investigated the influence of wind shear on the amplitude modulation of wind turbine noise in 2015. Wind shear and atmospheric turbulence effects on wind turbine noise using the Monin–Obukhov similarity theory was studied in 2016 by Yuan and Cotté [21]. To predict the trailing-edge noise, a more advanced trailing-edge noise model (TNO) was developed by Parchen [22] using a relation between the sound pressure level at far field and the surface pressure spectrum at trailing edge and related the far field sound spectrum as a function of turbulent boundary layer quantities. A refined TNO model using CFD was made by Kamruzzaman et al. [23] to consider the non-isotropic issue in the trailing-edge boundary layer. Tian [24] extended Amiet's model for simulating noise from a wind turbine.

This paper deals with the development of an advanced fluid-structure-noise technique for simulating noise generated from a wind turbine and controlling the noise generation of the wind turbine in wind shear and yaw by adjusting the pitch angle of its blades. A coupled flow and acoustics code that simultaneously predicts the noise and aerodynamic outputs of a wind turbine is applied in this study.

The paper is organized as follows. Section 2 presents the related theory and methodology. Validation and control results are presented in Section 3. Finally, conclusions are drawn in Section 4.

2. Theory and Methodology

The flow past a wind turbine is modelled with the actuator line (AL) method introduced by Sørensen and Shen [25], which was implemented in DTU's in-house finite volume code EllipSys3D. The actuator line method was coupled successfully with a high-order spectral method by Kleusberg et al. [26,27]. EllipSys3D was developed by the co-operation of the Department of Mechanical Engineering at DTU [28] and the Department of Wind Energy at Risø National Laboratory [29] that is now the Department of Wind Energy at DTU. The AL method [25,30] represents the aerodynamic loads of wind turbine blades by using a body force distributed along rotating lines. The body force imposed in the Navier–Stokes equations is computed by the aero-elastic code FLEX5 [31] also developed at DTU. The coupling of FLEX5 and EllipSys3D/AL methods is performed at every time-step. The flow data at the blades from EllipSys3D/AL are interpolated from the flow mesh and then fed into FLEX5. Using these information, the angle of attack and relative velocity including blade deformation/motions are calculated in FLEX5 and the aerodynamic loads on the blades are obtained further by using tabulated airfoil lift, drag and moment vs. angle of attack (AoA) data including dynamic stall. The new blade positions and loads from FLEX5, which includes the effect of blade bending and motion of the rotor (up/down/side-to-side/rotations) are then fed into EllipSys3D/AL as a body force on the moving actuator lines. With these new blade positions, the new loads are redistributed on the CFD mesh and a new solution is obtained in EllipSys3D/AL. In a standard aero-elastic code, these effects and motions are calculated by using the blade momentum theory (BEM) as the aerodynamic model. Here, the high-fidelity Navier–Stokes AL method was used to solve the fluid-structure interaction problem instead. To perform with large eddy simulation, the mixed scale model of Ta Phuoc [32,33] was used. To calculate the noise emission from the blades, the BPM code was used. The variables of angle of attack and relative velocity were calculated in FLEX5 and fed to BPM. A flow chart of the combined EllipSys3D/AL, FLEX5 and BPM framework is shown in Figure 1.

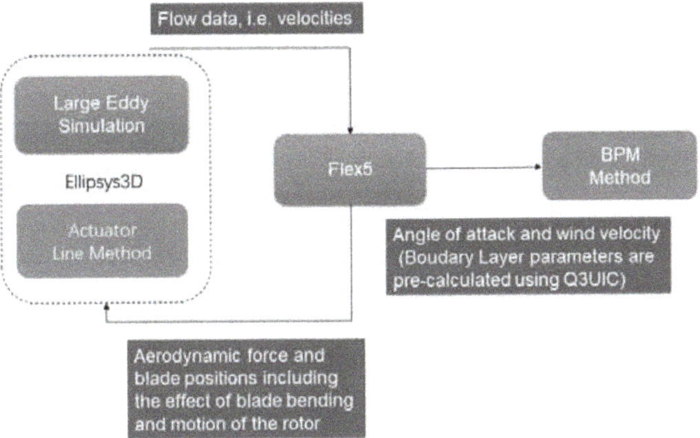

Figure 1. Flow chart of the combined EllipSys3D/AL, FLEX5 and BPM program.

In EllipSys3D, the solution to the incompressible Navier–Stokes equations is advanced in time using an iterative time-stepping method. The velocity-pressure coupling equations at each time-step are solved iteratively by sub-iterations with the usage of under-relaxation. First, the code solves the momentum equations as a predictor so that the solution can be advanced in time. The pressure correction equation, i.e., the rewritten continuity equation in satisfying the local mass flux conservation in the discretized form, is solved as a corrector making the final flow field satisfy the continuity constraint. This is a two-step procedure corresponding to a single sub-iteration, and the process is repeated until the solution becomes convergent within sub-iterations. The variables are then updated after the convergent solution is updated, and then followed by the next time step.

The aero-elastic code FLEX5 was designed to simulate a wind turbine's dynamic behavior regarding different wind conditions. FLEX5 operates in the time domain, and the output is the time-series of simulated loads and deflections. The detailed information of nine modules in FLEX5 can be found in [34].

The acoustic prediction BPM model used in this work takes the detailed geometrical and flow information into account, e.g., tip shape, blade geometry (chord and twist distributions) as well as instantaneous wind speed and direction. The velocity at each blade segment is computed by accounting for the induction effect and the vibration velocity of the blades. Furthermore, the geometry contours of airfoil sections of blade segments are also included in the prediction as important inputs, such as the boundary thickness and displacement thickness. The boundary layer parameters for a NACA 0012 was benchmarked with the viscous-inviscid interaction code XFOIL [35] and the strong viscous-inviscid interactive coupling code Q^{3}UIC [36], and compared with the measured values [13]. The comparisons (not shown) indicated that XFOIL under-predicts the boundary layer thicknesses about 30% and Q^{3}UIC gave a good agreement with the measurement. In this study, we compute the boundary quantities of airfoils using Q^{3}UIC. For more advanced calculations, the structure deformation in the blade torsion direction can also be considered, which will lead to changes in angle of attack. More details about the noise prediction calculation can be found in [18,37].

In wind turbine control systems, the proportional–integral–derivative (PID) controller is widely applied to control the generator speed and blade pitch angle. Obviously, the ideal controller should be developed by using the PID controller theory. In the present study, given the practical complexity, two control strategies, of (1) full step pitch input control and (2) interpolated pitch input control, are developed, which are P-type controls.

For the full step pitch input control, the blade pitch β is controlled at each time-step with an objective of AoA at one or more selected cross-sections, α_i, equal to its mean value in a cycle at one or more selected cross-sections, $\alpha_{i,mean}$, as follows

$$\beta = \frac{1}{N} \sum_{i=1,...,N} \alpha_i - \alpha_{i,mean}$$

where N is the number of selected cross-sections. After approximately 100 s, the solution becomes stable. Alternatively, the interpolation pitch input control can be used and Figure 2 illustrates the concepts. The idea is to control the blade pitch in the way that (i) at the peaks or troughs of one or more cross-sections, the blade pitch is controlled as the one in the full step pitch input control; (ii) at the azimuth positions in between the blade pitch is controlled by a pitch angle interpolated from the pitch angles at the peaks and troughs in (i) according to its azimuth position. Concretely, the peaks and troughs are extracted first with their corresponding azimuth angles and the mean AoAs are obtained from the maximums and minimums (red circles in Figure 2):

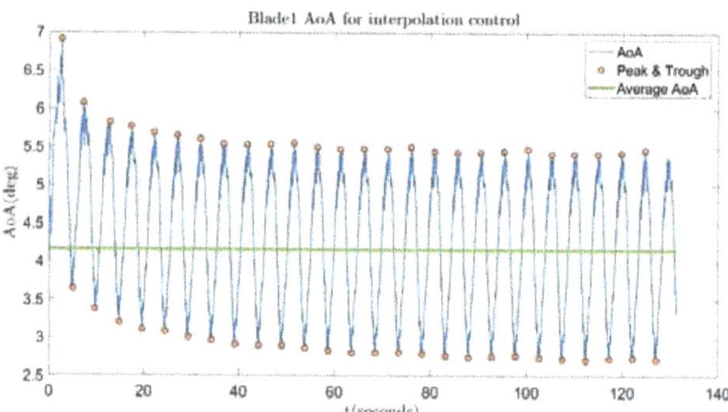

Figure 2. Pitch input setup based on angle of attack (AoA).

3. Numerical Results

In this section, the accuracy of the used numerical code is first assessed by performing a validation against field measurements performed in the DanAero project [38]. Results from the control study under the same conditions are presented afterwards.

3.1. Validation of the Simulation Framework

Before going further to discuss the performance of a wind turbine in different operation setups, the accuracy of the simulation code should be verified first. The 2.3 MW NM80 wind turbine was used here.

The design geometry of the LM 38.8 blade and aerodynamic airfoil data for the generic 2.3 MW variable speed and pitch controlled NM80 wind turbine were provided through a confidential agreement of using the DANAERO database. The airfoil data were used directly in AL/FLEX5, and the structural data originally prepared for HAWC2 were used to produce an equivalent FLEX5 input file. We assume the wind turbine operates in an ideal situation, so the tilt angle and tower shadow effects are neglected. Some key parameters of the wind turbine are listed in Table 1.

Table 1. Key wind turbine parameters.

Parameters	Value	Parameters	Value
Rotor diameter	80 m	Rated wind speed	16 m/s
Hub height	60 m	Max rotor speed	16.2 rpm
Rated power	2.3 MW		
Cut-in wind speed	4.5 m/s	Cut-out wind speed	25 m/s

Specific parameters for the two verification cases are listed below in Table 2.

Table 2. Description of Case I and Case II.

Parameters	Case I	Case II
Pitch angle	−4.75°	−4.75°
Rotor speed	16.2 rpm	16.2 rpm
Hub height wind speed	9.792 m/s	8.429 m/s
Shear exponent	0.249	0.262
Air density	1.22 kg/m^3	1.22 kg/m^3
Yaw angle	−6.02°	−38.34°
Ambient pressure	1005 Pa	1020 Pa

Figure 3 describes the configuration of the wind turbine and some parameters defined in Table 2.

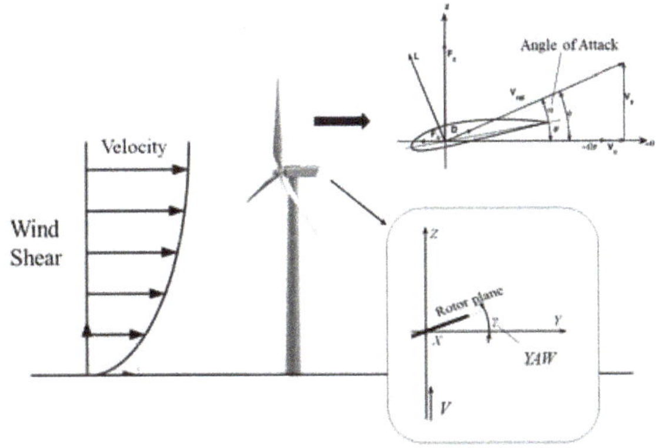

Figure 3. Configuration of a wind turbine under wind shear and yaw.

In the computations, a relatively fine mesh of about 19 million cells was used with $192 \times 192 \times 512$ cells in the transversal, vertical and streamwise directions in a domain of $[−16R, 16R] \times [0R, 19R] \times [−16R, 34R]$ (R is the rotor radius), respectively. A time-step based on rotor radius and inflow wind speed of 0.002 was used. A resolution of 30 cells per rotor radius was used in the rotor plane and 18 cells per rotor radius were uniformly distributed from 1D (rotor diameter) in front of the turbine to 11D behind the turbine in order to have a good resolution in the wake region. The airfoil data used in the computations were 2-dimensional airfoil data provided from the International Energy Agency (IEA) Task 29 consortium. To take into account the dynamic stall effects, the Øye dynamic stall model [34] was used as this is a part of FLEX5.

The boundary condition used in the computations is that at the inlet (min z), spanwise boundary (x direction) and up-boundary (max y), a sheared wind velocity profile is used, at the ground (y = 0) no-slip condition is used, and at the outlet (max z) the convective boundary condition is used. In this study, no synthetic inflow turbulence is used.

Figures 4 and 5 illustrate the comparisons between the measured data and simulation results. The simulation results were extracted from a cycle after the flow was stabilized. The forces normal and tangential to the local airfoil chord at 33%, 48%, 76% and 92% rotor radius noted as Fn33, Ft33, Fn48, Ft48, Fn76, Ft76, Fn92, Ft92, respectively, are shown below:

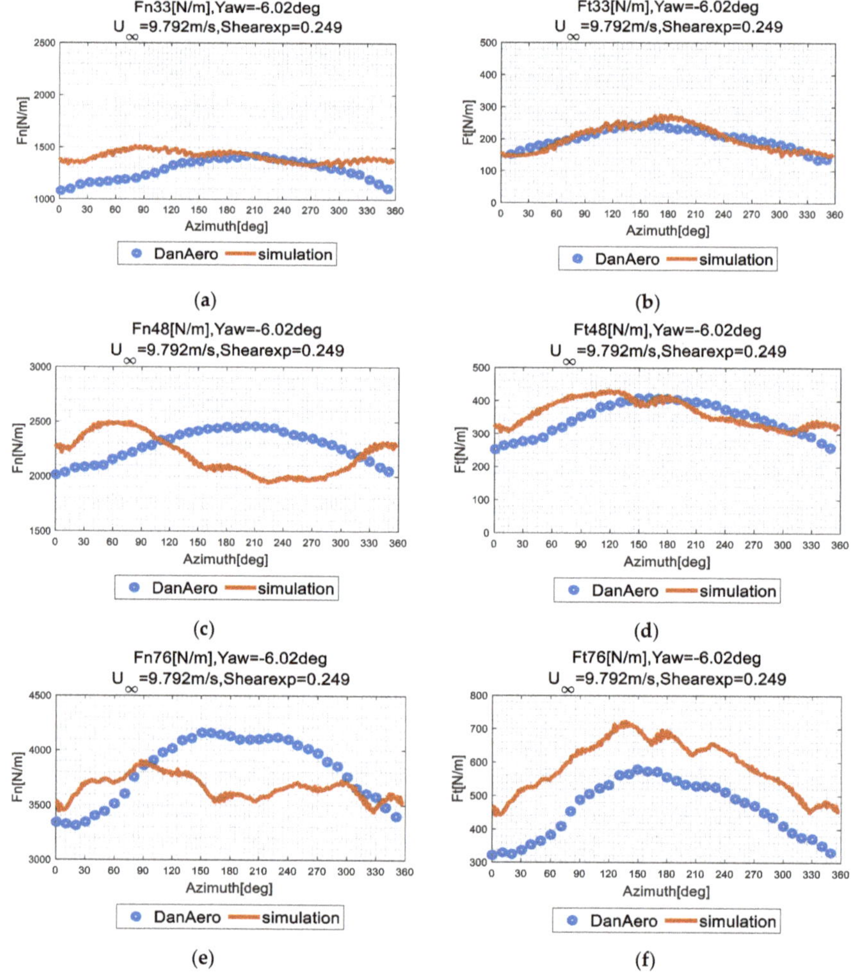

Figure 4. *Cont.*

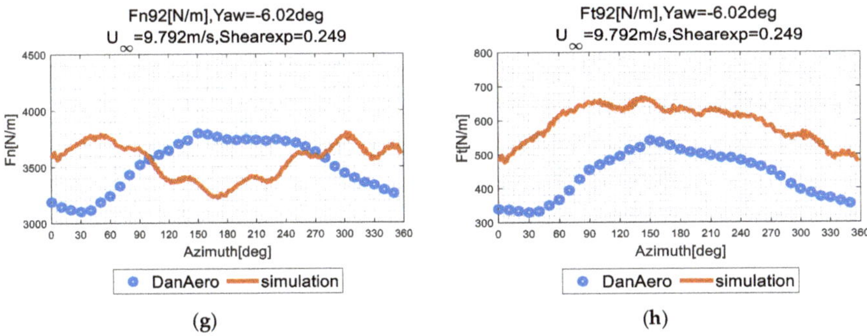

(**g**) (**h**)

Figure 4. Forces normal and tangential to the local chord at 33%, 48%, 76% and 92% rotor radius for Case I (**a**) Fn33; (**b**) Ft33; (**c**) Fn48; (**d**) Ft48; (**e**) Fn76; (**f**) Ft76; (**g**) Fn92; (**h**) Ft92.

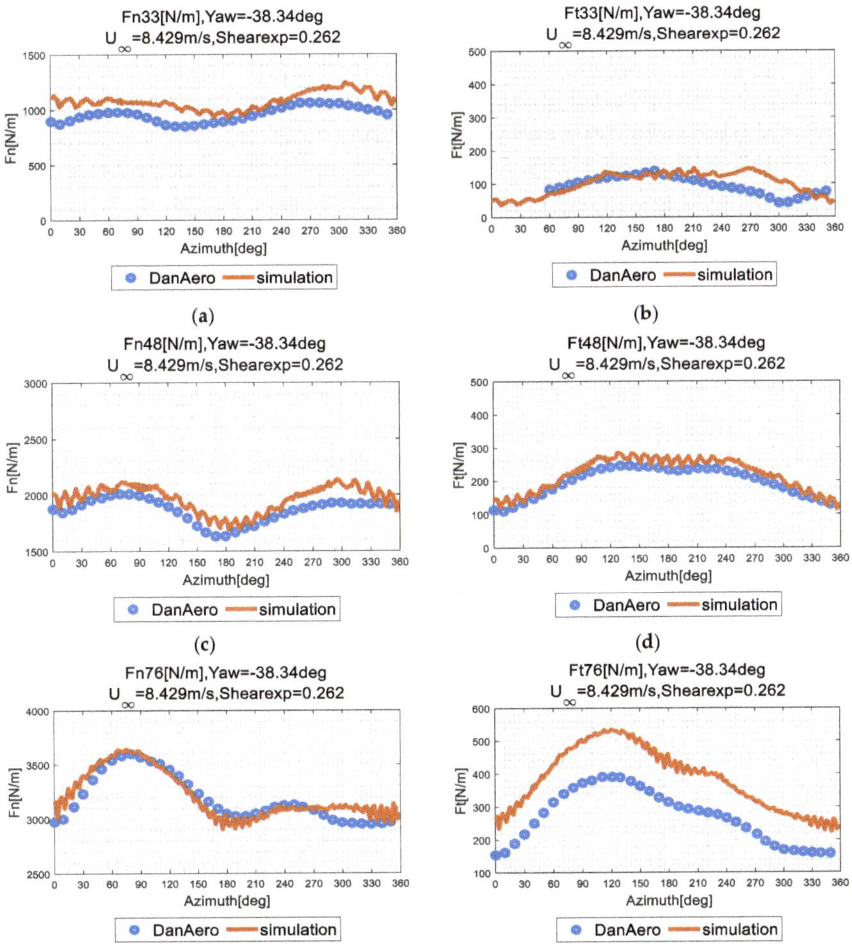

(**a**) (**b**)

(**c**) (**d**)

(**e**) (**f**)

Figure 5. *Cont.*

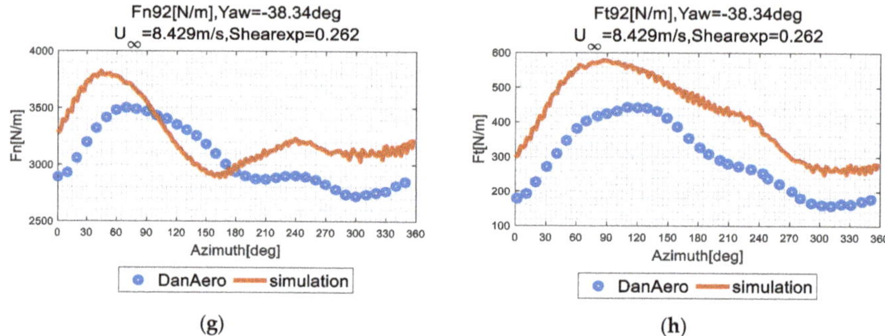

(**g**)　　　　　　　　　　(**h**)

Figure 5. Forces normal and tangential to the local chord at 33%, 48%, 76% and 92% rotor radius for Case II (**a**) Fn33; (**b**) Ft33; (**c**) Fn48; (**d**) Ft48; (**e**) Fn76; (**f**) Ft76; (**g**) Fn92; (**h**) Ft92.

As observed, the magnitudes between the measurements and simulations are similar for most of the quantities. In Case I, the computed normal force does not follow with the measured one and this is probably due to the Øye dynamic stall model used in the computations, which does not perform well at small angles of attack. On the other hand, the tangential force agrees well with the measurements. In Case II, the normal force is seen to be very well captured, while the tangential force is slightly over-predicted. The trends in function of azimuth angle (with 0° when the blade is down-pointing) followed the numerical results well. Since the tangential force is sensitive to the local angle of attack and airfoil data, calibrating airfoil data and its 3-dimensional corrections to rotational effects is a challenge for accurately predicting the performance of a wind turbine under wind shear and yaw. From the comparisons, it is also seen that the agreement is fairly good for both small (6°) and large (38°) yaw angles under a moderate wind shear. Therefore, it can be concluded from the validation that the simulation technique used in the present paper can predict the wind turbine performance relatively well.

3.2. Power Performance and Load Simulation

Since a wind turbine is rarely working with a large yaw angle, we define scenarios for the control study according to Case I in the validation. The specifics of the four different scenarios are described below:

Case 1:　No yaw, no shear;

Case 2:　10° yaw, no shear;

Case 3:　No yaw, 0.3 wind shear power law exponent;

Case 4:　10° yaw, 0.3 wind shear power law exponent.

Figure 6 illustrates the streamwise velocity and pressure field for Case 1 and Case 4. Horizontal axis x and vertical axis y are non-dimensional with the rotor radius of R = 40 m, which means the real values of them should equal to the axis values times R. The rotor center is located at x = 1.5 and y = 1.5. In Figure 6a, a velocity slowdown (induction effect) created by the rotor when exacting wind energy can be observed in the areas both in front of and behind the rotor, where the impact of induction effect is larger in the wake. When no shear is included, no wind speed stratification above the ground can be observed. A significant wind speed stratification above the ground can be observed in Figure 6b, while the wake is also mitigated by the lower longitudinal wind speed due to the shear. Similarly, Figure 6c,d illustrate the pressure field for Case 1 and Case 4, respectively. In general, the pressure in front of the rotor is larger than the pressure behind the rotor. By introducing the wind shear and yaw, the pressure both in front of and behind the rotor is changed, especially for the pressure in front of the rotor, which becomes more homogeneous.

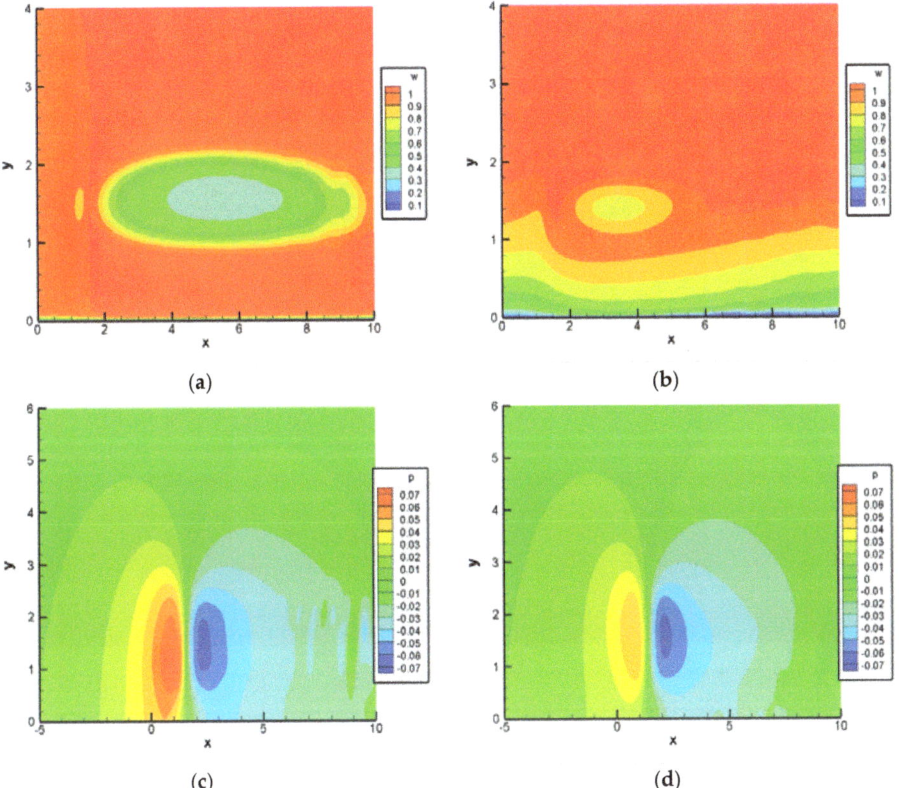

Figure 6. Streamwise velocity field and pressure field for Case 1 and Case 4: (**a**) streamwise velocity for Case 1; (**b**) streamwise velocity for Case 2; (**c**) pressure for Case 1; (**d**) pressure for Case 2.

The angle of attack (AoA) at 90% length of the blade for the four scenarios with no control is investigated. The comparison of the AoA can be found in Figure 7. In general, it can be observed that the fluctuations of AoA become much larger by introducing yaw or shear. In this study, the 0.3 wind shear power law exponent leads to a more significant amplitude change than the 10-degree yaw angle.

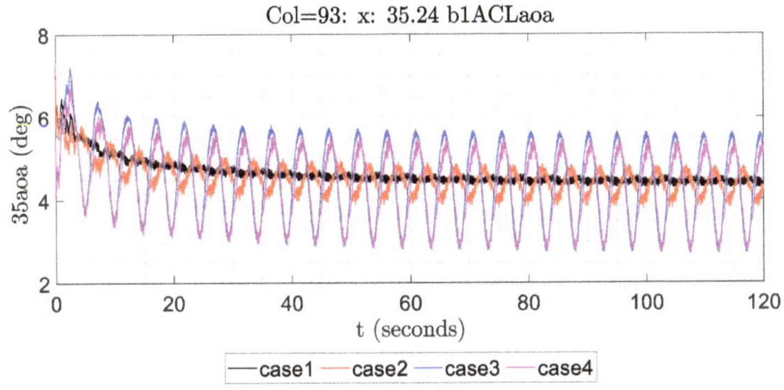

Figure 7. Angle of attack in a cross-section of 35.24 m (90%R) without control strategy.

Figure 8 illustrates the comparisons of AoA under pitch control with respect to different scenarios. It is noted that the pitch angle control in this study is performed using the AoA at 90%R where the wind turbine noise source center is located [37]. In the beginning, due to inexact interpolation values, the interpolated pitch control strategy results in more throbbing of AoA than the full step control strategy. However, when it is converged, both control techniques lead to a significant reduction of AoA fluctuations in all the four scenarios, while the mean of AoA remains the same.

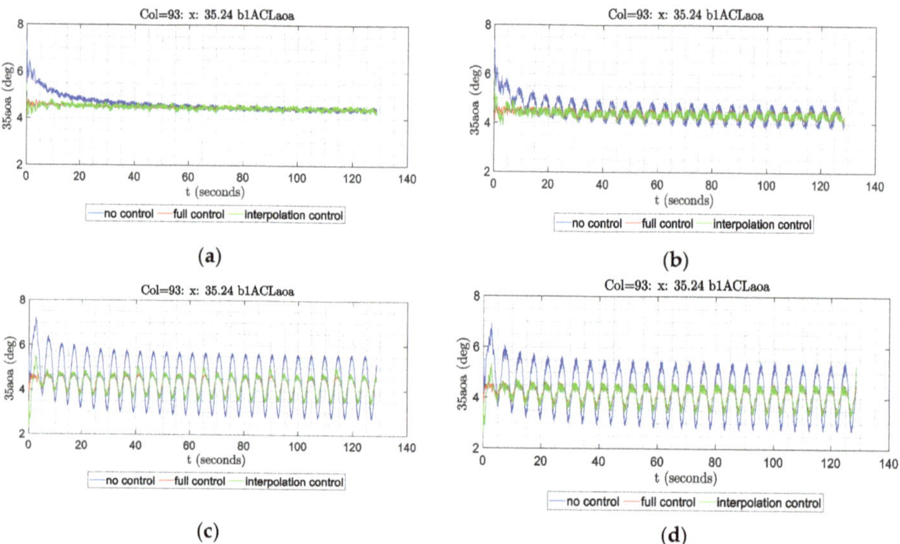

Figure 8. Angle of attack in section 35.24 m (90%R) with control strategy: (**a**) Case 1; (**b**) Case 2; (**c**) Case 3; (**d**) Case 4.

Figure 9 shows the pitch input values based on the results of AoA mentioned above. Pitch angles for two control strategies are calculated by using the angle of attack at 90%R of the no control case. It can be observed that the pitch is constant with 0.15° pitch offset when no control is performed. In terms of pitch in all the four cases, no significant difference can be found between control strategies I and II.

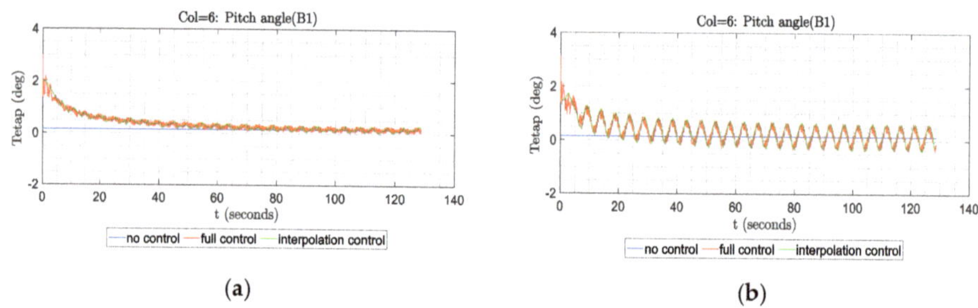

Figure 9. *Cont.*

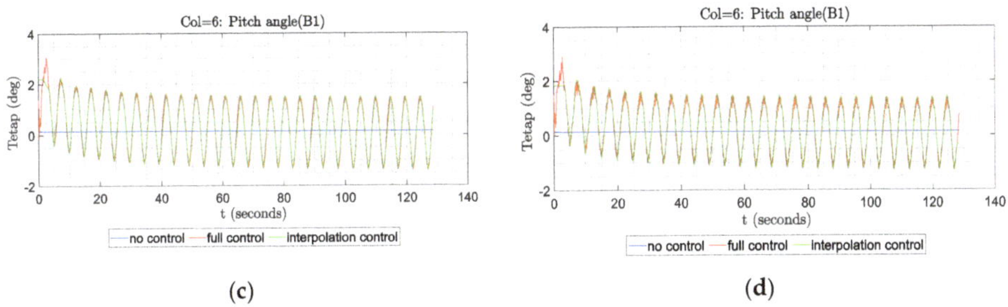

(c) (d)

Figure 9. Pitch in a cross-section of 35.24 m (90%R) with and without control strategy: (**a**) Case 1; (**b**) Case 2; (**c**) Case 3; (**d**) Case 4.

In Figure 10, power and thrust coefficients, noted as C_p and C_T, respectively, are plotted for the four scenarios. It can be seen that C_p is similar in all the four scenarios whereas introducing the shear and yaw slightly reduces C_p especially when C_p becoming more stationary, i.e., after 110 s. The reduction of the mean C_p is mainly due to the yaw misalignment, where the axial wind speed is smaller. It also should be noted that the scenarios with a yaw angle, C_p has more periodical fluctuations due to the yaw angle. A similar trend like C_p can be found in the thrust coefficient C_T, shown in Figure 10b, where the reduction of C_T is realized by the shear and yaw. The most significant difference happened in the scenario with both yaw and wind shear. Moreover, it can be observed that the difference between the yaw versus the yaw plus shear scenarios is relatively small. The reduction of C_T in Case 2, Case 3 and Case 4 is due to the reduction of the axial wind speed caused by the yaw.

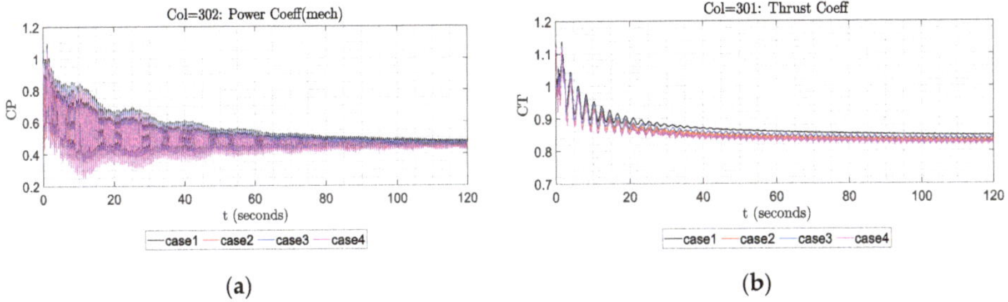

(a) (b)

Figure 10. C_p and C_T without control strategy: (**a**) power coefficient; (**b**) thrust coefficient.

By implementing the pitch adjustment, more fluctuations in Figure 11 are observed in the C_p and C_T signals while the mean C_p and C_T almost remain the same. By introducing the control methods, the fluctuation magnitudes of both control strategies increase, especially in control technique I. Some jumps of C_p and C_T values occur in the transient period as well. Moreover, the transient periods when performing both control techniques are longer, as compared to no control. For control technique I, the unexpected large fluctuations in C_T and C_p might be explained by the large pitch changes at each time step. For control technique II, similar results are observed when the fluctuation magnitudes are much smaller. The possible explanation is that there are fewer input pitch values and the corresponding interpolation enables better convergence of the elastic code.

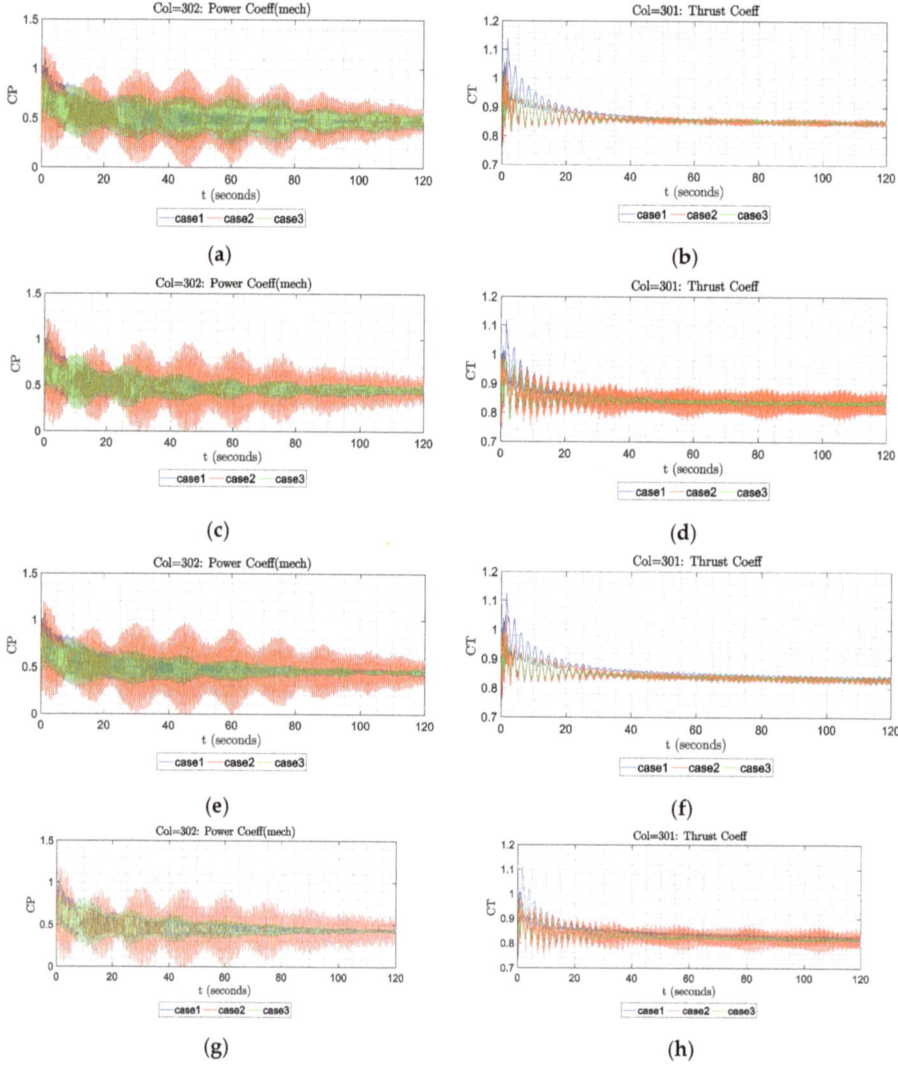

Figure 11. Power and thrust coefficients with a pitch adjustment: (**a**) C_p for Case 1; (**b**) C_T for Case 1; (**c**) C_p for Case 2; (**d**) C_T for Case 2; (**e**) C_p for Case 3; (**f**) C_T for Case 3; (**g**) C_p for Case 4; (**h**) C_T for Case 4.

3.3. Acoustical Analysis

Effects of Wind Shear and Yaw on Sound Emission

Figure 12 illustrates the polar plot of the sound pressure level in the function of azimuth angle of blade 1 for the four cases, which is called the amplitude modulation (AM) noise. AM noise is often a big issue because it gives more perception to the people living nearby. However, it should be noted that there are two different phenomena that make AM noise: one is caused by the blade noise directivity change during its operation [4] and another is caused by the non-uniform flow and operation, such as wind shear and yaw. In the present study, the latter is focused. The observer is located at 1.5 m height and a distance of 500 m downstream from the turbine (500 m away from the turbine is often a place where people are living although the sound propagation effects due to ground and atmospheric conditions are not considered here). The choice of the 500 m distance downstream location is to minimize the AM noise created by the blade noise directivity change. The speed of sound used in the calculations is 340 m/s. The turbulence level and turbulence length scale are set as 3% and 100 m, respectively. This choice was made in accordance with the 0.3 wind shear exponent defined in the scenarios. A more realistic choice can be made when the detailed flow information is measured. The 0° azimuth angle is defined when blade 1 is vertically down and coincides with the tower. All these four scenarios are performed with no control. As can be observed in Cases 1 and 3, the sound pressure level is slightly larger when the shear is included, while it significantly decreases with a 10° yaw, i.e., for both Cases 2 and 4. Normally, the SPL from one blade is highest when it is pointing upwards. The reason can be seen from the BPM sound prediction equations that the SPL is related to the fifth power of the oncoming velocity. For the case of three blades, at zero azimuth angle (blade 1 is pointing downwards), the other two blades are pointing at 120° and 240° when the noise is almost highest during its operation for Cases 3 and 4.

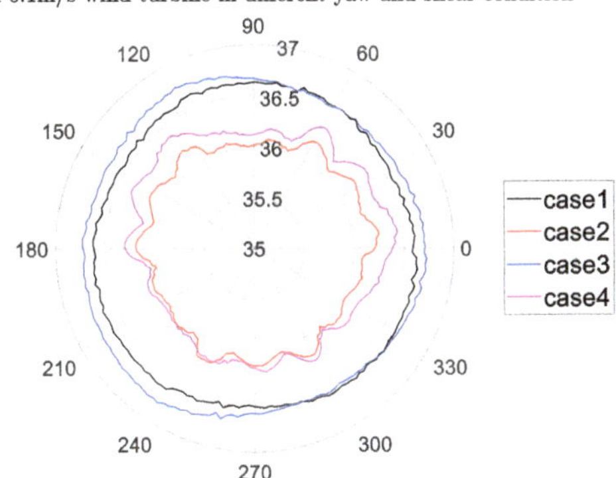

Figure 12. Sound pressure level in function of blade azimuth angle without control at 1.5 m height and 500 m distance downstream from the turbine at a wind speed of 6 m/s.

Figure 13 illustrates the comparison of SPL regarding different control scenarios. It can be observed that compared with the non-controlled case, the controlled SPL results, i.e., with control techniques I and II, are similar for Case 1 and Case 2. For Case 1, the peaks for both control strategies are in step, whereas the magnitude of the no control results is slightly smaller. For Case 2, the fluctuations are also generally in step, and both the time and magnitude of the peaks are consistent. Therefore, the implementation of the control techniques may not lead to noise reductions for the no-shear scenarios. The main reason for the negligible difference in Cases 1 and 2 is that the modifications of the pitch and the corresponding AoA are relatively small where the fluctuations of noise level are mainly due to blade deflections. A larger difference is expected for a larger pitch. It should be noted that Case 2 corresponds to a slightly lower SPL as compared to Case 1, due to the yaw-caused lower axial wind speed. For Cases 3 and 4, a larger decrease of SPL, i.e., about 0.4 dB, can be observed by introducing either control strategy I or II, whereas the trend remains the same. No significant difference can be found between the results of the two control techniques. It can be concluded that by introducing the control techniques, the reduction of SPL can be realized for the shear scenarios. In comparison to the angle of attack plots in Figure 8 with those plots of SPL in Figure 13, a non-linear effect between these two quantities is seen due to the blade structural responses. The AoA variation of the interpolation control in Figure 8d is reduced about 1°. According to the rule of a 1° AoA reduction resulting in a 1 dB noise reduction [39], the noise change in the present study is within this range. For comparison, it can be observed that the differences between the controlled and uncontrolled results of A-weighted SPL regarding all the cases are negligible. The reason is that the

A-weighted SPL accounts for more of the high-frequency noise to which humans are more sensitive and de-emphasizes the low-frequency contribution. In our case, the reduction of the noise mainly occurs in the low-frequency part and thus the A-weighting SPL is similar in all the cases. It should be noted that the low-frequency noise is important in far-field as it can propagate in a long distance with very small air absorption. To check the AM noise, the modulation depth for Case 4 is calculated according to the method proposed in [7] and shown in Table 3. From the table, it is seen that the modulation depth is slightly reduced.

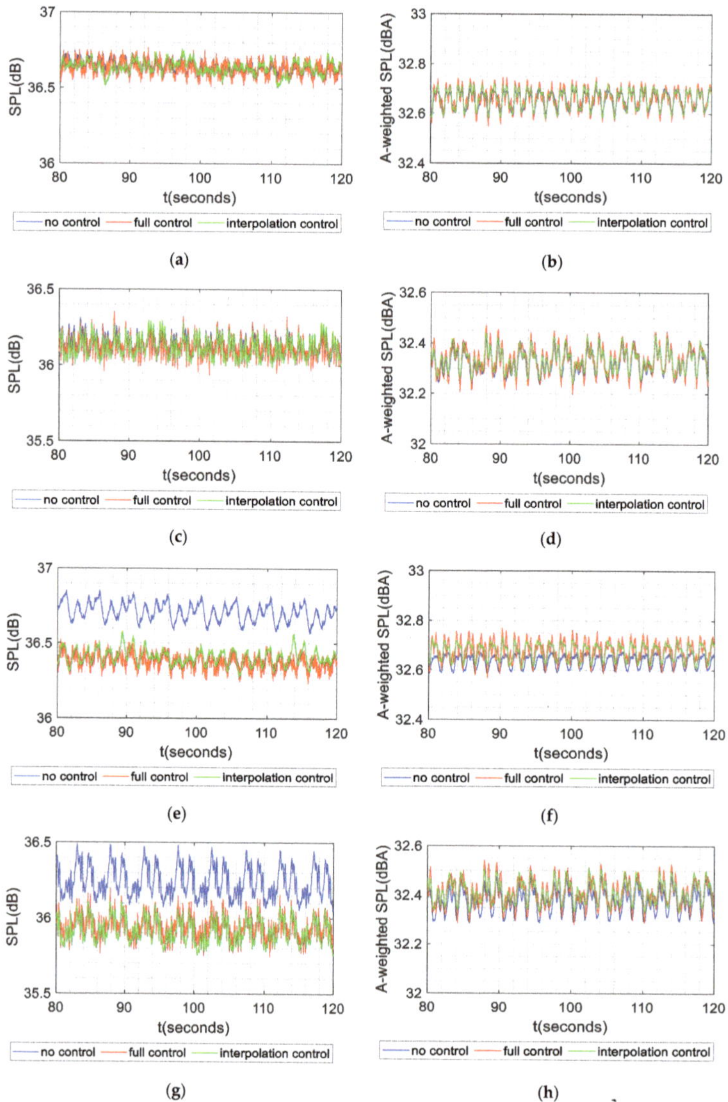

Figure 13. Sound pressure level and A-weighted sound pressure level of the turbine at 6 m/s wind speed: (**a**) sound pressure level (SPL) for Case 1; (**b**) A-weighted SPL for Case 1; (**c**) SPL for Case 2; (**d**) A-weighted SPL for Case 2; (**e**) SPL for Case 3; (**f**) A-weighted SPL for Case 3; (**g**) SPL for Case 4; (**h**) A-weighted SPL for Case 4.

Table 3. Modulation depth for Case 4.

condition	6 m/s no control	6 m/s full control	6 m/s inter control
modulation depth (dB)	0.3331	0.3127	0.2736
condition	10 m/s no control	10 m/s full control	10 m/s inter control
modulation depth (dB)	0.4308	0.3834	0.3692

Moreover, this phenomenon can be verified in Figure 14, where the contribution of SPL in different frequencies at the time instant (120 s) is shown. From the figure, it is seen that at the low frequency (which is less than 315 Hz), SPL is reduced obviously by the pitch control strategy.

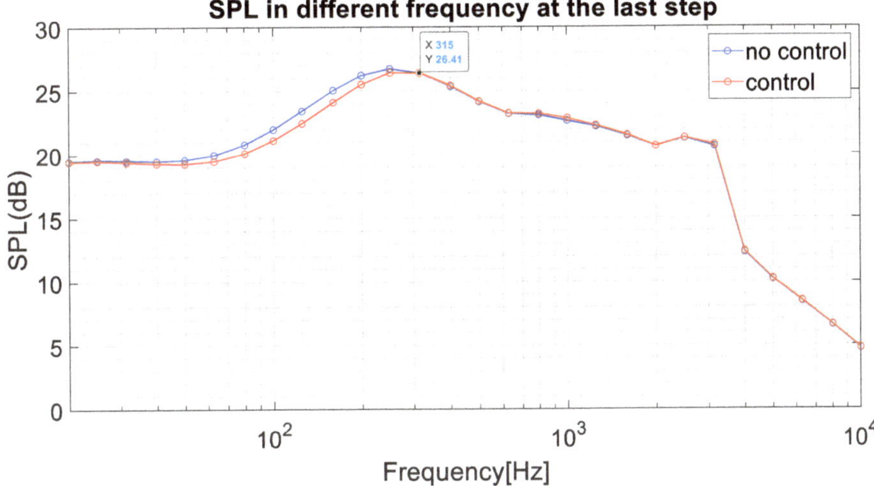

Figure 14. SPL spectra of case 4 of the turbine at 6 m/s wind speed at a time instant of 120 s after control.

Figure 15 illustrates the SPL and A-weighted SPL in Cases 1, 2 and 4, but with a lager inflow wind speed of 10 m/s. With the lager wind speed, the SPL increases about 8.5 dB in each case. The interesting thing deserved to be noticed is that the pitch control methods at 10 m/s display a similar effect in Case 4 as at 6 m/s but the variations in amplitude are more important. Since the A-weighted SPL was designed according to human hearing and emphasizes the contribution from the high frequency sound centered at 1000 Hz, it can give an impression that the control methods give a small effect on A-weighted SPL. The modulation depth is listed in Table 3 and a small reduction is seen. It is worth noting that the yaw and shear considered in this study are relatively small and these effects can be much bigger in large shear and yaw cases. Moreover, the noise propagation effects due to changing atmospheres and wakes [8] are not included here. In the future, the present noise source-modelling framework will be further coupled with the propagation model [8] in order to control wind turbine noise at far field.

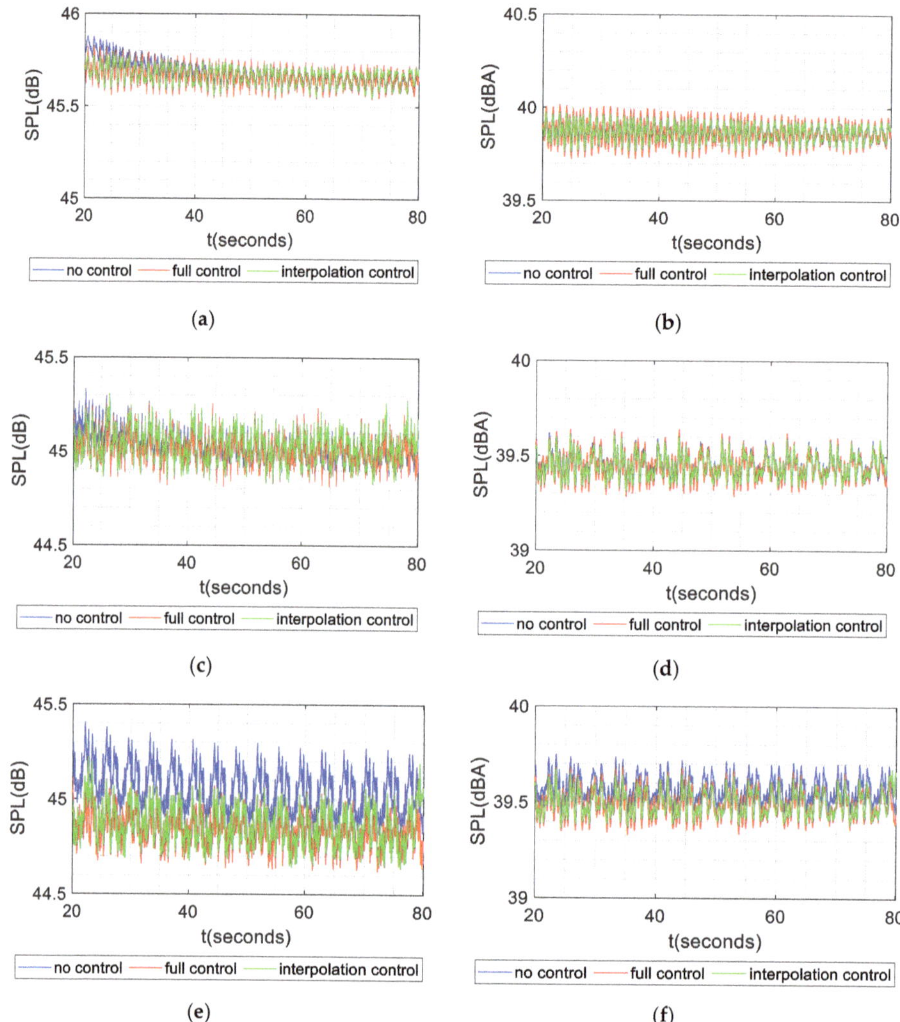

Figure 15. Sound pressure level and A-weighted Sound pressure level from the wind turbine at 10 m/s wind speed: (**a**) SPL for Case 1; (**b**) A-weighted SPL for Case 1; (**c**) SPL for Case 2; (**d**) A-weighted SPL for Case 2; (**e**) SPL for Case 4; (**f**) A-weighted SPL for Case 4.

4. Conclusions

In this paper, an advanced flow-structure-acoustics framework has been developed and validated against load field measurements for the case of the 2.3 MW NM80 wind turbine under a moderate wind shear and a small and a large yaw angle. The code has been further used for studying and controlling noise generation from a wind turbine under a moderate wind shear and a small yaw, which is often the case when a wind turbine operates. Through analysis of the results, several conclusions are summarized here:

1. Simulations based on the coupled Ellipsys3D/AL/Flex5 framework agreed well with field load measurements on the 2.3 MW NM80 turbine.

2. It was observed that the wind shear had a significant impact on the flow field, which led to a fainter wake. The longitudinal wind speed was reduced by introducing yaw, which accordingly decreased the sound pressure level.

3. The reduction of sound pressure level under a moderate wind shear situation with a shear exponent of 0.3 was undertaken by a pitch adjustment, while the power and thrust coefficients generally became more undulant. The pitch adjustment techniques were only useful for the case with a wind shear.

4. Given the miscellaneous reasons causing the wind turbine noise and the complex relations in SPL prediction formulations, it was obviously found that SPL and AoA were not linearly related. This non-linear relationship was also confirmed by comparing the variation of AoA and SPL.

5. Through the comparison of SPL and A-weighted SPL, the pitch control implemented in this paper mainly affected the wind turbine noise at low frequencies, while unapparent effects occurred with high-frequency noise.

6. The mean SPL was reduced with 0.4 dB and the modulation depth was reduced slightly.

Author Contributions: Conceptualization, W.Z.S.; Formal analysis, M.Z.; Investigation, M.Z.; Methodology, W.Z.S.; Project administration, W.Z.S.; Resources, W.Z.S.; Software, M.S. and W.Z.S.; Supervision, M.S., H.Y. and W.Z.S.; Visualization, M.Z.; Writing—original draft, M.Z.; Writing—review & editing, M.S., H.Y. and W.Z.S. All authors have read and agreed to the published version of the manuscript.

Funding: This paper was partially funded by the Danish Energy Agency in the project (EUDP2018, Journal nr. 64018-0084) and the Innovation Fund Denmark in the project entitled "DecoWind—Development of low-noise and cost-effective wind farm control technology" under project number (8055-00041B). The DanAero projects were funded partly by the Danish Energy Agency, (EFP2007. Journal nr. 33033-0074 and EUDP 2009-II. Journal nr: 64009-0258) and partly by self-funding from the project partners Vestas, Siemens Gamesa Renewable Energy, LM Wind Power, Dong Energy and DTU Wind Energy.

Conflicts of Interest: The authors declare no conflict of interest.

Abbreviations

BPM	a semi-empirical noise prediction model developed by Brooks, Pope and Marcolini
AL	actuator line
BEM	blade momentum theory
AoA	angle of attack
SPL	sound pressure level
PID	proportional–integral–derivative controller

References

1. Van den Berg, F.; Pedersen, E.; Bouma, J.; Bakker, R. *Project WINDFARM Perception: Visual Acoustic Impact of Wind Turbine Farms on Residents—Final Report. Report FP6–2005-Science-and-Society-20*; University Medical Centre Groningen: Groningen, Netherlands, 2008.

2. *Danish Regulation for Wind Turbine Noise*; Ministry of Environment and Food of Denmark: Copenhagen, Denmark, 2019. Available online: https://mst.dk/service/nyheder/nyhedsarkiv/2019/feb/ny-bekendtgoerelse-om-stoej-fra-vindmoeller/ (accessed on 13 February 2020).

3. How Loud Is a Wind Turbine. Available online: https://www.windpowerengineering.com/loud-wind-turbine-really/ (accessed on 27 October 2020).

4. Oerlemans, S. Detection of Aeroacoustic Sound Sources on Aircraft and Wind Turbines. Ph.D. Thesis, University of Twente, Enschede, the Netherlands, 2009.

5. Fukushima, A.; Yamamoto, K.; Uchida, H.; Sueoka, S.; Kobayashi, T.; Tachibana, H. Study on the amplitude modulation of wind turbine noise: Part 1—Physical investigation. In Proceedings of the INTER-NOISE, Innsbruck, Austria, 15–18 September 2013.

6. Coles, D.; Levet, T.; Cand, M. Application of the UK IOA Method for Rating Amplitude Modulation. In Proceedings of the 7th International Conference on Wind Turbine Noise, Rotterdam, the Netherlands, 2–5 May 2017.

7. Fernandez, F.A.; Burdisso, R.A.; Arenas, J.P. Indoor simulation of amplitude modulated wind turbine noise. *Wind Energy* **2017**, *20*, 507–519. [CrossRef]

8. Barlas, E.; Zhu, W.J.; Shen, W.Z.; Kelly, M.; Andersen, S.J. Effects of wind turbine wake on atmospheric sound propagation. *Appl. Acoust.* **2017**, *122*, 51–61. [CrossRef]

9. Schlinker, R.; Amiet, R. Helicopter rotor trailing edge noise. In Proceedings of the 7th Aeroacoustics Conference, Palo Alto, CA, USA, 5–7 October 1981; NASA CR-3470. pp. 1–145. [CrossRef]

10. Grosveld, F.W. Prediction of Broadband Noise from Horizontal Axis Wind Turbines. *J. Propuls. Power* **1985**, *1*, 292–299. [CrossRef]

11. De Wolf, W.B. EEN Predictiemethode voor het Aerodynamische Geluid van Windturbines met Horizontale AS. In *NLR TR 870181L*; Netherlands Aerospace Centre: Amsterdam, the Netherlands, 1986.

12. Viterna, L.A. The NASA-LeRC Wind turbine sound prediction code. In *NASA CP-2185*; NASA Lewis Research Center: Cleveland, OH, USA, 1981.

13. Brooks, T.F.; Pope, D.S.; Marcolini, M.A. Airfoil self-noise and prediction. In *NASA RP-1218*; NASA Langley Research Center: Hampton, VA, USA, 1989.

14. Fuglsang, P.; Madsen, H.A. Implementation and verification of an aeroacoustic noise prediction model for wind turbines. In *Risø-R-867(EN)*; Risø National Laboratory: Roskilde, Denmark, 1996.

15. Lowson, M.V. Assessment and prediction of wind turbine noise. In *Flow Solutions Report 92/19—ETSU W/13/00284/REP*; Harwell Laboratory, Energy Technology Support Unit: Harwell, UK, 1992.

16. Amiet, R.K. Acoustic radiation from an airfoil in a turbulent stream. *J. Sound Vibr.* **1975**, *41*, 407–420. [CrossRef]

17. Moriarty, P.; Migliore, P. Semi-empirical aeroacoustic noise prediction code for wind turbines. In *NREL/TP-500-34478*; National Renewable Energy Lab.: Golden, CO, USA, 2003; pp. 1–39.

18. Zhu, W.J.; Heilskov, N.; Shen, W.Z.; Sørensen, J.N. Modeling of aerodynamically generated noise from wind turbines. *J. Solar Energy Eng.* **2005**, *127*, 517–528. [CrossRef]

19. Bowdler, D. Wind shear and its effect on noise assessment. In Proceedings of the Third International Conference on Wind Turbine Noise, Aalborg, Denmark, 17–19 June 2009. Available online: https://www.dickbowdler.co.uk/content/publications/Wind-Shear-and-its-Effect-on-Noise-Assessment.pdf (accessed on 26 October 2020).

20. Oerlemans, S. Effect of wind shear on amplitude modulation of wind turbine noise. *Int. J. Aeroacoust.* **2015**, *14*, 715–728. [CrossRef]

21. Tian, Y.; Cotté, B. Wind turbine noise modeling based on Amiet's theory: Effects of wind shear and atmospheric turbulence. *Acta Acust. United Acust.* **2016**, *102*, 626–639. [CrossRef]

22. Parchen, R. Progress report DRAW, a prediction scheme for trailing-edge noise based on detailed boundary-layer characteristics. In *TNO-Report HAG-RPT-980023, Report*; TNO Institute of Applied Physics: The Hague, the Netherlands, 1998.

23. Kamruzzaman, M.; Lutz, T.; Würz, W.; Shen, W.Z.; Zhu, W.J.; Hansen, M.; Bertagnolio, F.; Madsen, H.A. Validations and improvements of aeroacoustics models using detailed experimental data. *Wind Energy* **2012**, *15*, 45–61. [CrossRef]

24. Tian, Y. Modeling of Wind Turbine Noise Sources and Propagation in the Atmosphere. Ph.D. Thesis, ENSTA Paris Tech, Palaiseau, France, 2016.

25. Sørensen, J.N.; Shen, W.Z. Numerical modeling of wind turbine wakes. *J. Fluids Eng.* **2002**, *124*, 393–399. [CrossRef]

26. Kleusberg, E.; Mikkelsen, R.F.; Schlatter, P.; Ivanell1, S.; Henningson, D.S. High-order numerical simulations of wind turbine wakes. *J. Phys. Conf. Ser.* **2017**, *854*, 012025. [CrossRef]

27. Kleusberg, E. Wind-Turbine Wakes—Effects of Yaw, Shear and Turbine Interaction. Ph.D. Thesis, Royal Institute of Technology, Stockholm, Sweden, 2019.

28. Michelsen, J.A. Basis3D-a platform for development of multiblock PDE solvers. In *Technical Report AFM 92-05*; Technical University of Denmark: Copenhagen, Denmark, 1992.

29. Sørensen, N.N. General purpose flow solver applied to flow over hills. In *Risø-R-827(EN)*; Risø National Lab.: Roskilde, Denmark, 1995.

30. Shen, W.Z.; Zhu, W.J.; Sørensen, J.N. Actuator line/Navier–Stokes computations for the MEXICO rotor: Comparison with detailed measurements. *Wind Energy* **2012**, *15*, 811–825. [CrossRef]

31. Øye, S. FLEX Simulation of wind turbine dynamics. In *State of the Art of Aeroelastic Codes for Wind Turbine Calculations*; Pedersen, M.B., Ed.; The Technical University of Denmark: Lyngby, Denmark, 1996.

32. Ta Phuoc, L. Modèles de Sous Maille Appliqués aux Ecoulements Instationnaires Décollés. In *Aérodynamique Instationnaire Turbulente-Aspects Numériques et Expérimentaux*; DGA/DRET: Paris, France, 1994.

33. Shen, W.Z.; Zhu, W.J.; Sørensen, J.N. Aero-acoustic computations for turbulent airfoil flows. *AIAA J.* **2009**, *47*, 1518–1527. [CrossRef]

34. Øye, S. *FLEX5 Manual*; Department of Wind Energy, Technical University of Denmark: Lyngby, Denmark, 1999.

35. Drela, M. XFOIL, an analysis and design system for low Reynolds number airfoils. In Proceedings of the Conference on Low Reynolds Number Aerodynamics, University of Notre Dame, Notre Dame, IN, USA, 5–7 June 1989; pp. 1–12.

36. Garcia, N.R.; Sørensen, J.N.; Shen, W.Z. A strong viscous-inviscid interaction model for rotating airfoils. *Wind Energy* **2014**, *17*, 1957–1982. [CrossRef]

37. Leloudas, G.; Zhu, W.J.; Sørensen, J.N.; Shen, W.Z.; Hjort, S. Prediction and reduction of noise from a 2.3 MW wind turbine. *J. Phys. Conf. Ser.* **2007**, *75*, 012083. [CrossRef]

38. Madsen, H.A.; Bak, C.; Troldborg, N. The DanAero experiments—A detailed investigation of MW wind turbine aerodynamics and aeroacoutics. *Wind Energy* **2020**. submitted.

39. Wagner, S.; Bareib, R.; Guidati, G. *Wind Turbine Noise*, 1st ed.; Springer: Berlin/Heidelberg, Germany, 1996.

 *applied sciences*

Article

Performance Characteristics of an Orthopter-Type Vertical Axis Wind Turbine in Shear Flows

Rudi Purwo Wijayanto [1,2,*], Takaaki Kono [3,*] and Takahiro Kiwata [3]

[1] Graduate School of Natural Science and Technology, Kanazawa University, Kanazawa 920-1192, Japan
[2] The Agency of the Assessment and Application of Technology (BPPT), Jakarta 10340, Indonesia
[3] Institute of Science and Engineering, Kanazawa University, Kanazawa 920-1192, Japan; kiwata@se.kanazawa-u.ac.jp
* Correspondence: rudi.purwo@stu.kanazawa-u.ac.jp (R.P.W.); t-kono@se.kanazawa-u.ac.jp (T.K.)

Received: 9 January 2020; Accepted: 2 March 2020; Published: 4 March 2020

Abstract: To properly conduct a micro-siting of an orthopter-type vertical axis wind turbine (O-VAWT) in the built environment, this study investigated the effects of horizontal shear flow on the power performance characteristics of an O-VAWT by performing wind tunnel experiments and computational fluid dynamics (CFD) simulations. A uniform flow and two types of shear flow (advancing side faster shear flow (ASF-SF) and retreating side faster shear flow (RSF-SF)) were employed as the approaching flow to the O-VAWT. The ASF-SF had a higher velocity on the advancing side of the rotor. The RSF-SF had a higher velocity on the retreating side of the rotor. For each type of shear flow, three shear strengths (Γ = 0.28, 0.40 and 0.51) were set. In the ASF-SF cases, the power coefficients (C_p) were significantly higher than the uniform flow case at all tip speed ratios (λ) and increased with Γ. In the RSF-SF cases, C_p increased with Γ. However, when Γ = 0.28, the C_P was lower than the uniform flow case at all λ. When Γ = 0.51, the C_P was higher than the uniform flow case except at low λ; however, it was lower than the ASF-SF case with Γ = 0.28. The causes of the features of C_P were discussed through the analysis of the variation of blade torque coefficient, its rotor-revolution component and its blade-rotation component with azimuthal angle by using the CFD results for flow fields (i.e., horizontal velocity vectors, pressure and vorticity). These results indicate that a location where ASF-SFs with high Γ values dominantly occur is ideal for installing the O-VAWT.

Keywords: orthopter; vertical axis wind turbine; power coefficient; torque coefficient; shear flow; wind tunnel; CFD; delayed detached-eddy simulation

---

1. Introduction

Since the 2000s, interest in installing small wind turbines (SWTs) in the built environment has been growing [1–9]. Wind conditions in the built environment are complex in nature and are characterized by lower wind speeds and higher turbulence because of the presence of obstructions [8,9]. For SWTs to be able to make up their costs within their lifetimes, they should have high efficiency and be placed at sites with high wind speeds, such as coastal sites or high-elevation inland sites. However, in the built environment, keeping the rotational speed of an SWT's rotor as low as possible is preferable from the viewpoint of aerodynamic noise [10,11]. Therefore, the optimal tip-speed ratio of an SWT in the built environment should be as low as possible, while the maximum power coefficient of the SWT should be as high as possible.

Wind turbines are classified into horizontal-axis wind turbines (HAWTs) and vertical-axis wind turbines (VAWTs), based on the orientation of their rotation axes. Generally, in the built environment, VAWTs are preferable to HAWTs because VAWTs do not suffer, as much as HAWTs, from reduced energy outputs from frequent wind direction changes [12]. Wind turbines are further classified into lift-type wind turbines and drag-type wind turbines, based on the aerodynamic force component that acts on a blade and dominantly contributes to the rotor rotation. With regard to lift-type VAWTs, a lot of research on Darrieus-type VAWTs, including straight-bladed and helical-bladed ones, has been conducted [13–16]. With regard to drag-type VAWTs, a lot of research on Savonius-type VAWTs has been conducted [17–21]. In general, the optimal tip-speed ratio of a drag-type VAWT is less than 1.0 [11], which is much smaller than that of a lift-type VAWT. Moreover, although the maximum power coefficient of a drag-type VAWT is generally much smaller than that of a lift-type VAWT, the power coefficient of a drag-type VAWT is generally greater than that of a lift-type VAWT at a low tip-speed ratio, of less than 1.0. Therefore, a drag-type VAWT is favorable in the built environment and was researched by our research group.

Our group [22,23] researched a drag-type VAWT called the orthopter-type VAWT (O-VAWT). The O-VAWT is a variable-pitch VAWT; each of the flat-plate blades not only revolves around the main shaft but also rotates around its own blade axis, which is rotationally supported by a pair of connecting arms. We investigated the effects of the number and aspect ratio of the flat-plate blades on the power performance of the O-VAWT in a uniform flow by conducting wind tunnel experiments with an open test section and three-dimensional computational fluid dynamics (CFD) simulations. When the number of the blades was three and the aspect ratio of the blades was 1:1, the maximum power coefficient was 0.25 and the optimal tip-speed ratio was 0.4 [22]. Here, the optimal tip-speed ratios of Savonius-type VAWTs are in the range of 0.45 to 1.0 [21]. Except for several studies that were conducted using a wind tunnel with a very high blocking ratio of the closed test section, the maximum power coefficient of the Savonius-type VAWT was, at most, 0.25 [21]. That is, the O-VAWT has a lower optimal tip-speed ratio than Savonius-type VAWTs, although the maximum power coefficient is relatively high. Therefore, from the viewpoint of aerodynamic noise, the O-VAWT can be more favorable in the built environment as compared to Savonius-type VAWTs. Except for our studies, studies on the power performance of O-VAWTs are very limited. Shimizu et al. [24] investigated the effects of the aspect ratio of the blade on the power performance of an O-VAWT with two blades whose cross-sectional shape was an ellipse by conducting wind tunnel experiments. Bayeul-Line et al. [25] examined the effects of the blade's cross-sectional shape (elliptical and straight) and the initial blade stagger angle on the performance of an O-VAWT by conducting 2-dimensional CFD simulations. Cooper and Kennedy [26] examined the power performance of an O-VAWT with three blades whose cross-sectional shape was the upstream half of a NACA0010-65 section reflected about the mid-chord by conducting theoretical analysis with a multiple-stream tube model and field measurements. Our group [23] conducted wind tunnel experiments to compare the performance of an O-VAWT with elliptic blades and one with flat-plate blades. By considering the mechanical loss torque, we obtained the maximum power coefficient of 0.246 at a tip-speed ratio of 0.4 for the O-VAWT with elliptic blades and of 0.288 at a tip-speed ratio of 0.4 for the O-VAWT with flat-plate blades. It should be noted that these studies on O-VAWTs were conducted in conditions where the approaching flows had uniform distribution.

To properly conduct a micro-sitting of an O-VAWT in the built environment, it is important to understand the effects of the strong shear approach flow with on the performance of the O-VAWT. Figure 1 illustrates the approaching wind flow to a building. As the wind flow approaches the building, the wind speed decreases, and the pressure increases. Then, the wind flow proceeds along the upwind face of the building and separates at the corners on the roof and the side walls. As the separated wind flow is not obstructed by the building, the pressure decreases, and the wind speed increases. Near the upwind corners, the wind speed increases more than that of the approaching wind. In addition, reverse flow regions are formed between the separated shear layer and the building's walls. As a result, strong shear flows are formed vertically over the roof surface and horizontally over the side walls. Due to the mixing of momentum, the shear becomes weaker as the flow proceeds downstream. To utilize the increased wind speed over the roof of a building, the effects of building shapes and wind directions on the wind conditions have been investigated (e.g., [27]). Furthermore, the effects of wind conditions, such as wind speed, turbulent intensity, and skew angles, on the potential energy yield and the power performance of a wind turbine have been studied (e.g., [28,29]). In this study, we investigated the effects of horizontal shear flow on the performance of the O-VAWT by conducting wind tunnel experiments and three-dimensional CFD simulations. A uniform flow and two types of shear flow were employed as the approaching flow to the O-VAWT. One type had a higher velocity on the advancing side of the rotor. The other type had a higher velocity on the retreating side of the rotor. For each type of shear flow, we set three different shear strengths.

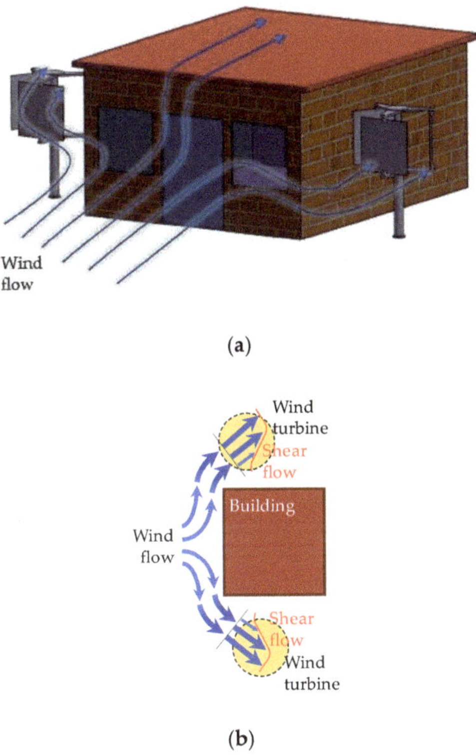

Figure 1. Illustration of the approaching wind flow to the wind turbine near upwind corners of a building. (**a**) Bird view; (**b**) enlarged top view.

2. Experimental Approach

2.1. Wind Turbine Model

In this paper, we employed a right-handed Cartesian coordinate system $(x_1, x_2, x_3) = (x, y, z)$, in which the z-direction was aligned in the vertical direction. The wind turbine used in this study was an O-VAWT with three flat-plate blades as shown in Figure 2. The blade had a height of $h = 4.00 \times 10^{-1}$ m, chord length of $c = 4.00 \times 10^{-1}$ m and thickness of 4.0×10^{-3} m. Each of the blades not only revolved around the main shaft but also rotated around its own blade axis, which was rotationally supported by a pair of connecting arms. The distance between the main shaft and one of the blade axes was $R = 2.55 \times 10^{-1}$ m. In addition, each of the blade axes was connected with the main shaft by a chain via sprockets. Since the ratio of the number of teeth on the sprocket of the main shaft to that of the blade axis was 1:2, each of the blades rotated around the own blade axis a half time while the rotor revolved around the main shaft one time. When seen from the top, the rotor revolved around the main shaft counterclockwise and each of the blades rotated around the own blade axis clockwise as shown in Figure 2c. Therefore, by using the angular velocity of the rotor revolution (ω), the angular velocity of the blade rotation is expressed as $-\omega/2$. Furthermore, as shown in Figure 2c, according to the azimuthal angle of a blade (φ), we call the range of $90° < \varphi < 270°$ the "upwind region" of the rotor, $0° < \varphi < 90°$ and $270° < \varphi < 360°$ the "downwind region" of the rotor, $180° < \varphi < 360°$ the "advancing side of the rotor" and $0° < \varphi < 180°$ the "retreating side" of the rotor. The O-VAWT is designed so that the drag force on a blade is large on the advancing side of the rotor while being small on the retreating side of the rotor.

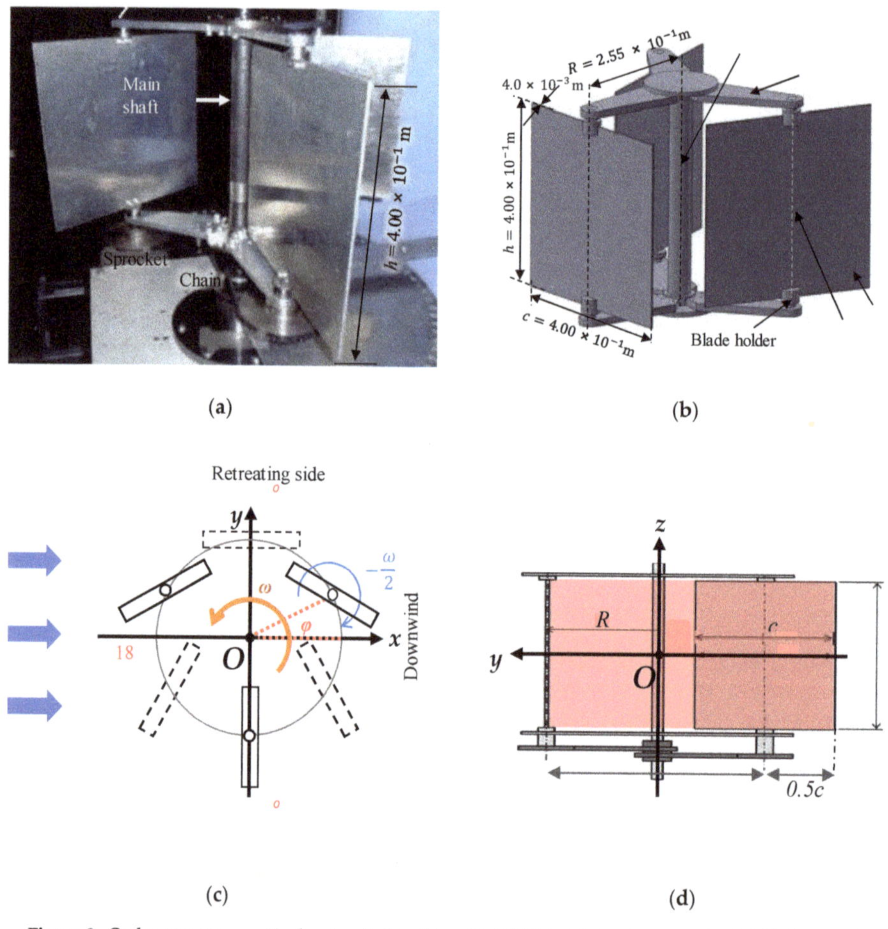

Figure 2. Orthopter-type vertical-axis wind turbine (O-VAWT). (**a**) A photograph of O-VAWT with three flat blades, main shaft, arm, the chains and sprockets; (**b**) an isometric view of O-VAWT with three flat blades; (**c**) motion of rotor and blades viewed from the top; (**d**) a projected swept area of the rotor viewed from the upwind side.

Figure 2d shows a projected swept area of the rotor (A), which is defined as:

$$A = (2R + 0.5\,c)\,h. \tag{1}$$

We define the diameter of the rotor as:

$$D = 2R. \tag{2}$$

The O-VAWT had a rotor's diameter of $D = 5.1 \times 10^{-1}$ m and a projected rotor's swept area of $A = 2.84 \times 10^{-1}$ m^2 can be considered as a micro wind turbine. A small scale wind turbine that has a diameter up to 1.25 m and the swept area up to 1.2 m^2 is categorized as a micro wind turbine [4].

2.2. Experimental Setup for Uniform Flow Case

Figure 3a shows the experimental setup for the uniform flow case. The experiments were conducted using a closed circuit wind tunnel with an open test section. The size of the cross section of the wind tunnel outlet was 1.25 m × 1.25 m. The blockage ratio which is defined as the ratio of the projected rotor's swept area to the wind tunnel outlet area was approximately 18%. The O-VAWT was set in the test section so that the rotor center was at the center of the cross-section of the wind tunnel outlet and 0.850 m downwind of the wind tunnel outlet. Here, we defined the rotor center as the point on the rotational axis of the rotor and at the mid-height of the blades. In addition, we set the origin of the coordinate system at the rotor center, as shown in Figure 2c,d. The rotor was driven by a motor (Mitsubishi Electric, GM-S) and its rotational speed (ω) was monitored by using a digital tachometer (Ono Sokki, HT-5500) and controlled by using an inverter (Hitachi, SJ200). The rotor torque was measured by using a torque meter (TEAC, TQ-AR), which was connected to the motor and shaft via couplings. The output signal of the torque meter was converted by a 16-bit analog-to-digital converter with a sampling interval of 0.5°, and 36,000 items (50 revolutions) of data were stored. To measure the reference wind speed U_∞, an ultrasonic anemometer (Kaijo Sonic, DA-650-3TH and TR-90 AH) was set approximately 2 m upwind of the wind tunnel outlet. The value of U_∞ was kept at 8 m/s. The value of tip speed ratio λ, which is defined as:

$$\lambda = R\omega/U_\infty, \tag{3}$$

was varied from 0.1 to 0.8 with an increment of 0.1.

2.3. Experimental Setup for Shear Flow Cases

To generate a horizontal shear flow, a perforated panel and a splitter plate were installed at the outlet of the wind tunnel as shown in Figure 3b,c and Figure 4. The perforated panel had a width of 1.0 m, a height of 1.5 m and a thickness of 2×10^{-3} m, and was set so that it covered half of the wind tunnel outlet in the horizontal wind direction. Due to the existence of the perforated panel, the pressure upwind of the panel increased and the wind flow rate through the wind-tunnel-outlet area covered by the panel decreased while that through the uncovered wind-tunnel-outlet area increased. When the perforated panel covered the wind-tunnel-outlet area upwind of the retreating side of the rotor, the wind speed of the generated shear flow was higher on the advancing side of the rotor. Hereafter, this type of shear flow is referred to as "advancing side faster shear flow" (ASF-SF). On the other hand, when the perforated panel covered the wind-tunnel-outlet area upwind of the advancing side of the rotor, the wind speed of the generated shear flow was higher on the retreating side of the rotor. Hereafter, this type of shear flow is referred to as "retreating side faster shear flow" (RSF-SF). The splitter plate was set vertically, parallel to the wind tunnel wall and at the center of the wind tunnel outlet to avoid the horizontal component of the wind velocity in the generated shear flow becoming significant. The splitter plate had a width of 0.88 m, a height of 1.25 m and a thickness of 4×10^{-3} m.

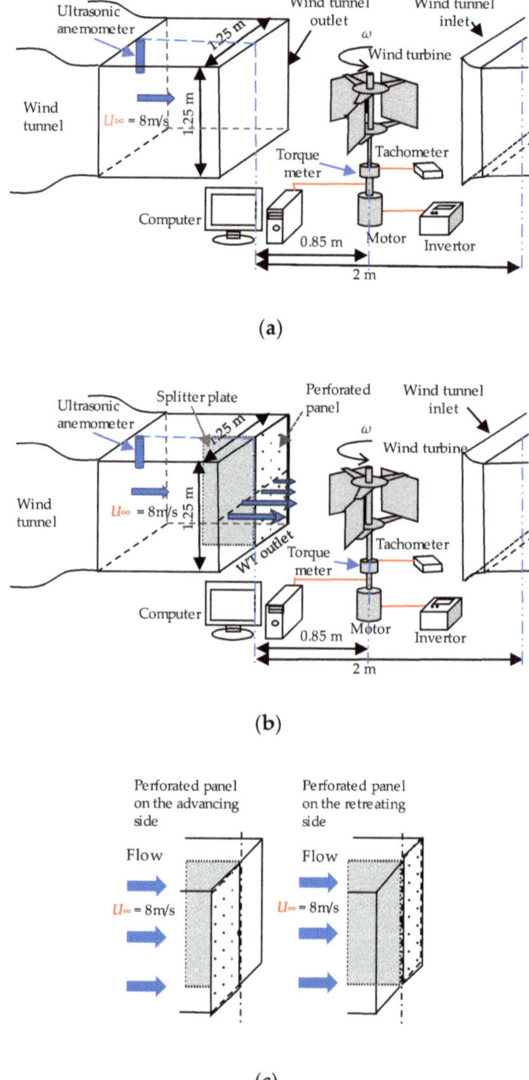

Figure 3. The experimental apparatus and measurement devices; (**a**) in uniform flow, (**b**) in shear flows and (**c**) the porous plate position at the nozzle exit of the wind tunnel in case of shear flows.

Figure 4. The perforated panel and splitter plate set at the outlet of the wind tunnel.

To investigate the effects of the strength of the shear flow on the performance of the O-VAWT, we generated three kinds of shear flows by using three perforated panels shown in Table 1. With regard to a staggered round-hole perforated panel, the shielding ratio Φ, which is the ratio of the area that shields the airflow to the whole area of the perforated panel, can be computed by:

$$\Phi = 1 - \frac{\pi d^2}{2\sqrt{3}\,L^2},\qquad(4)$$

where d is a diameter of a hole and L is the distance between the centers of adjacent holes.

Table 1. Perforated panels. Here, d is the diameter of a hole, L is the distance between the centers of adjacent holes and Φ is the shielding ratio.

Name	d [m]	L [m]	Φ [-]	Enlarged View
Perforated panel A	3×10^{-3}	4×10^{-3}	0.49	4mm
Perforated panel B	3×10^{-3}	4.5×10^{-3}	0.60	4.5mm
Perforated panel C	3×10^{-3}	5×10^{-3}	0.67	5mm

Except for the installation of the perforated panels and the splitter plate, the experimental setup for the measurement of the performance of the O-VAWT was the same as the uniform flow case. Prior to the measurement of the performance of the O-VAWT, we measured the horizontal profiles of the generated shear flows at 0.10 m downwind of and at the center height of the wind tunnel outlet by using an x-type hot-wire probe (Kanomax, 0252R-T5).

2.4. Torque and Power Coefficients

Due to the difficulty of evaluating the mechanical losses of the bearings, the sprockets and the chains, this study considers only the aerodynamic torque generated by the blades as the rotor torque of the O-VAWT. The aerodynamic torque generated by the blades was computed by:

$$T_B = T_{wB} - T_{woB},$$ (5)

where T_{wB} was the measured aerodynamic torque generated by the rotor when the blades were not removed; and T_{woB} was the measured aerodynamic torque generated by the rotor when the blades were removed. It is worth noting that, generally, when the blades were not removed from the rotor, the rotor generated positive torque while the motor acted as a load to keep the value of ω constant. Conversely, at all tip speed ratios, when the blades were removed from the rotor, the rotor generated negative torque while the motor acted as the driving force of the rotor revolution. Therefore, at all tip speed ratios, T_B was higher than T_{wB}.

The power coefficient describes that fraction of the power in the wind that may be converted by the turbine into mechanical work [30] and is defined in this study as:

$$C_P = \frac{T_B \omega}{0.5 \rho A U_0^3},$$ (6)

and the torque coefficient is defined as:

$$C_T = \frac{T_B}{0.5\,\rho A U_0^2 R}.$$ (7)

Here, the U_0 is the time-mean stream-wise velocity, $\bar{u}(x, y, z)$, averaged over the projected rotor's swept area at $x = 0$ and is computed by:

$$U_0 = \frac{\int_{-(R+0.5c)}^{R} \bar{u}(0, y, 0)\,dy}{2R+0.5c}.$$ (8)

3. Numerical Approach

The CFD software utilized to simulate the wind flow field was ANSYS Fluent 17.2 [31,32]. The numerical approach was based on our previous paper [22,33], in which the CFD simulations with the delayed detached eddy simulation (DDES) turbulence model of flow around the O-VAWT were conducted and the validities of the grid resolution and the time-step size were confirmed.

3.1. Governing Equations and Discretization Method

The flow field around the wind turbine was assumed to be incompressible and isothermal. The DDES turbulence mode treats near-wall region in a manner like a Reynolds-averaged Navier–Stokes (RANS) turbulence model and treats the rest of the flow field in a manner like a large-eddy simulation (LES) turbulence model [34]. This model has the potential to achieve higher accuracy than RANS models and save a large number of computing resources compared with pure LES models. The governing equations for the CFD simulation with the DDES turbulence model based on the Spalart–Allmaras (SA) model are the continuity equation:

$$\frac{\partial u_i}{\partial x_i} = 0,\tag{9}$$

the Navier–Stokes equation:

$$\frac{\partial u_i}{\partial t} + \frac{\partial u_i u_j}{\partial x_j} = -\frac{1}{\rho}\frac{\partial p}{\partial x_i} + v\frac{\partial}{\partial x_j}\left(\frac{\partial u_i}{\partial x_j} + \frac{\partial u_j}{\partial x_i} - \frac{2}{3}\delta_{ij}\frac{\partial u_i}{\partial x_j}\right) - \frac{\partial}{\partial x_j}\left[\tilde{v}\left(\frac{\partial u_i}{\partial x_j} + \frac{\partial u_j}{\partial x_i}\right) - \frac{2}{3}k\delta_{ij}\right],\tag{10}$$

and the transport equation for the kinematic eddy viscosity $\tilde{v}$:

$$\frac{\partial \tilde{v}}{\partial t} + \frac{\partial \tilde{v} u_i}{\partial x_i} = C_{b1}\tilde{v}\left(S + \frac{\tilde{v}}{\kappa^2 d_{DDES}^2}\left(1 - \frac{\chi}{1 + \chi f_{v1}}\right)\right) + \frac{1}{\sigma\tilde{v}}\left[\frac{\partial}{\partial x_j}\left\{(v + \tilde{v})\frac{\partial \tilde{v}}{\partial x_j}\right\} + C_{b2}\left(\frac{\partial \tilde{v}}{\partial x_j}\right)^2\right] - C_{w1}f_w\left(\frac{\tilde{v}}{d_{DDES}}\right)^2\tag{11}$$

where u_i is the wind-velocity component in the x_i direction; p is the pressure; v is the kinematic viscosity; t is the time; ρ is the air density; k is the turbulence kinetic energy; δ_{ij} is the Kronecker delta; d_{DDES} is the DDES length scale; χ is ($\tilde{v}/v$); S is a scalar measure of the deformation tensor; f_{v1} and f_w are damping functions; and C_{b1}, C_{b2}, C_{w1}, $\sigma_{\tilde{v}}$ and κ are constants.

The DDES length scale is computed by:

$$d_{DDES} = d - f_d max\left(0, d - c_{des}\Delta_{max}\right)\tag{12}$$

where d is the distance to the closest wall; c_{des} is the empirical constant; and Δ_{max} is the maximum edge length of the local computational cell, i.e., $\Delta_{max} = max\left(\Delta_x, \Delta_y, \Delta_z\right)$. The switching between the RANS and the LES mode depends on the following shielding function;

$$f_d = 1 - tanh\left((8r_d)^3\right)\tag{13}$$

and

$$r_d = \frac{\tilde{v}}{\sqrt{u_{i,j}u_{i,j}\kappa^2 d^2}}\tag{14}$$

where $u_{i,j}$ is the velocity gradient. The damping functions and closure coefficients are as follows:

$$f_{v1} = \frac{\chi^3}{c_{v1}^3 + \chi^3}, \; f_w = g\left[\frac{1 + c_{w3}^6}{g^6 + c_{w3}^6}\right]^{1/6}, \; g = r + c_{w2}(r^6 - r), \; c_{w1} = \frac{c_{b1}}{\kappa^2} + \frac{(1 + c_{b2})}{\sigma_{\tilde{v}}},\tag{15}$$

$$c_{b1} = 0.1355, \; c_{b2} = 0.622, \; c_{v1} = 7.1, \; c_{w2} = 0.3, \; c_{w3} = 2.0, \; \sigma_{\tilde{v}} = \frac{2}{3}, \; c_{des} = 0.65, \; \kappa = 0.4187.\tag{16}$$

The governing equations are discretized by the finite-volume method. The advection terms of the Navier–Stokes equations are discretized by the bounded central-difference scheme. The advection term of the transportation equations for $\tilde{v}$ is discretized by a second-order upwind scheme. Other spatial derivatives are discretized by the central-difference scheme. The time integration is performed using the second-order implicit method.

3.2. Numerical Setup

The computational domain, the computational meshes and the boundary conditions are shown in Figure 5. As the same as the experimental setup, the origin of the coordinate system is defined at the center of the O-VAWT. The modeled O-VAWT was comprised of three blades, one main shaft and two sets of connecting arms. To reduce the computational cost, other components, such as the chains and sprockets are omitted. The sizes of these components are the same as those used in the experiment. The computational domain consists of three blade domains, one rotor domain and one far-field domain. The blade domain included one of the blades and rotates around each blade axis. The rotor domain included these three blade domains, the main shaft and the connecting arms and rotates around the main shaft. The far-field domain was a stationary domain and its size was $23.5D \times 17D \times 5D$. Except for uniform flow cases, a splitter plate with the same thickness as the experiment was set at $y = 0$ and $x = -11.37D$ to $-1.67D$. In all domains, only unstructured meshes were used. Based on our previous mesh-resolution dependency tests [33], the number of computational cells in each domain were set as shown in Table 2. The total number of computational cells was approximately 10 million. All surfaces of the solid components were covered with boundary-layer meshes. The first grid nodes over the surface of the blades were $y+ < 1$ in all run cases.

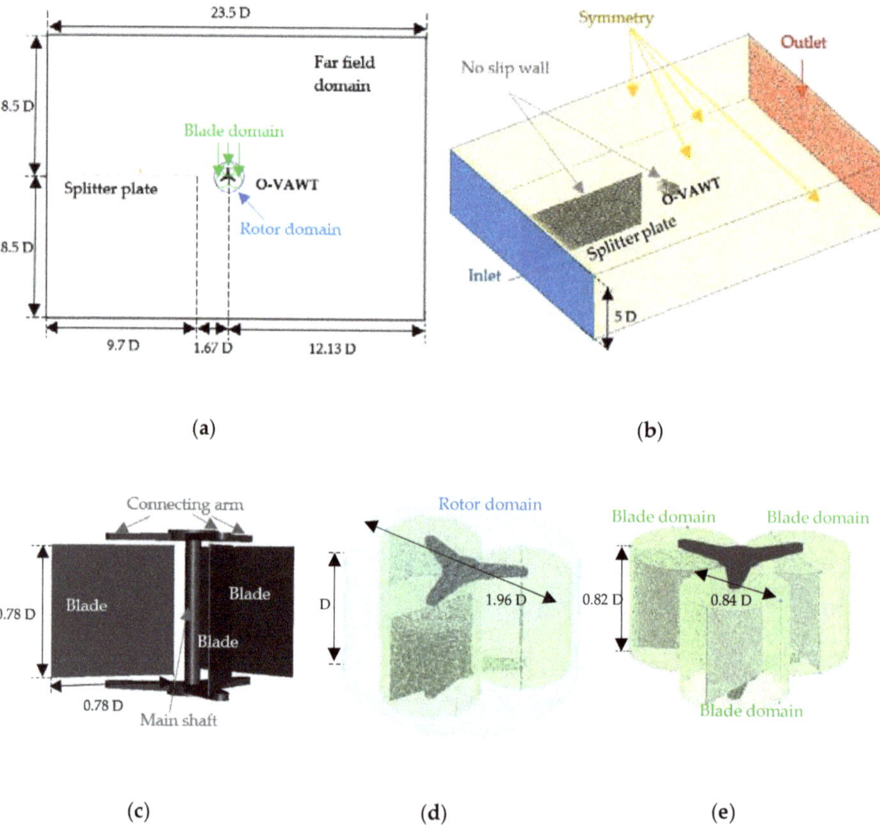

Figure 5. Computational domain, computational mesh and boundary conditions. (**a**) Top view of the computational domain; (**b**) bird's eye view of the computational domain with computational mesh and boundary conditions; (**c**) modeled rotor; (**d**) blade domains and (**e**) rotor domain.

Table 2. Numbers of grid sizes.

Domain	Mesh on This Research	Regular Mesh Used by ElCheikh [33]
Blade 1	2,300,530	1,610,660
Blade 2	2,303,725	1,614,903
Blade 3	2,304,607	1,588,363
Blade 4	-	1,614,661
Rotor	1,756,635	2,127,497
Far end	1,770,583	441,196
TOTAL	**10,436,080**	**8,997,280**

At the inlet boundary, the distributions of the stream-wise wind velocity shown in Table 3 were implemented. These distributions were set so that when the O-VAWT is absent, the distributions of the time-mean values of u at $x \approx -0.147\,D$, which corresponds to 0.1 m downwind of the wind tunnel outlet, matched well with those of the wind tunnel experiment, which were generated by using the three kinds of perforated panels with different shielding ratios Φ, as shown in Figure 6a,b. Here, U_H and U_L are the maximum and minimum streamwise velocities in the profile of a shear flow measured in the wind tunnel experiment; Γ is the velocity ratio defined as:

$$\Gamma = \frac{U_H - U_L}{U_H + U_L} \tag{17}$$

Table 3. Distribution of u at the inlet boundary. Here, Γ is the velocity ratio; Φ is the shielding ratio of a perforated panel; U_H and U_L are the maximum and minimum velocities in a shear flow.

Flow Type	Γ	Φ	U_H [m/s]	Range of U_H Region [m]	Velocity Distribution in Transition Region	Range of Transition Region [m]	U_L [m/s]	Range of U_L Region [m]
Uniform	0	-	8		8		8	
ASF-SF	0.51	0.67	13.2	$y \le -0.24$	$U_H \cdot (-y/0.24)^{0.3}$	$-0.24 \le y \le -0.01$	4.3	$y \ge -0.01$
ASF-SF	0.40	0.60	12.1	$y \le -0.22$	$U_H \cdot (-y/0.22)^{0.35}$	$-0.22 \le y \le 0.02$	5.2	$y \ge -0.02$
ASF-SF	0.28	0.49	10.8	$y \le -0.18$	$U_H \cdot (-y/0.18)^{0.2}$	$-0.18 \le y \le -0.01$	6.1	$y \ge -0.01$
RSF-SF	0.51	0.67	13.2	$y \ge 0.24$	$U_H \cdot (y/0.24)^{0.3}$	$0.01 \le y \le 0.24$	4.3	$y \le 0.01$
RSF-SF	0.40	0.60	12.1	$y \ge 0.22$	$U_H \cdot (y/0.24)^{0.35}$	$0.02 \le y \le 0.22$	5.2	$y \le 0.02$
RSF-SF	0.28	0.49	10.8	$y \ge 0.18$	$U_H \cdot (y/0.24)^{0.2}$	$0.01 \le y \le 0.18$	6.1	$y \le 0.01$

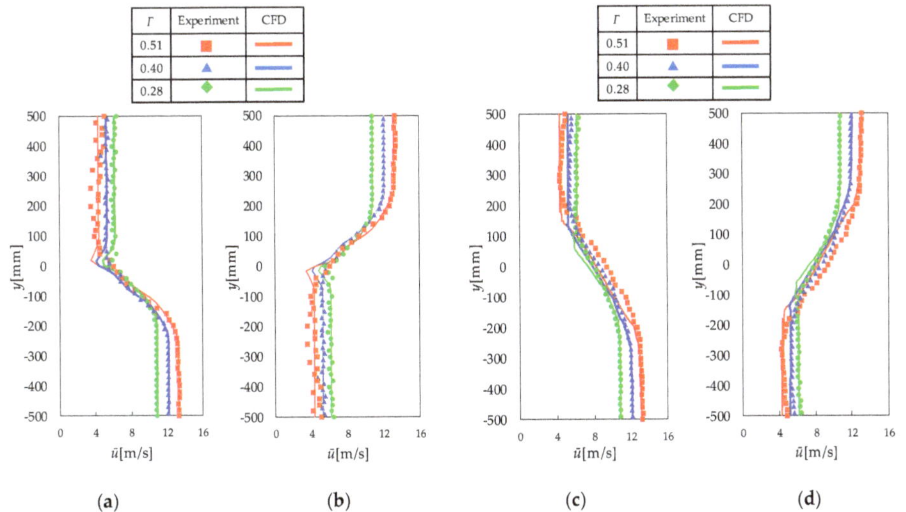

Figure 6. Horizontal distribution of the time-mean values of streamwise velocity at the mid-height of the wind tunnel outlet at $x \approx -0.147D$, which corresponds to 0.1 m downwind of the wind tunnel outlet, and $x = 0$, which corresponds to the position of the rotational axis of the O-VAWT, when the O-VAWT is absent. (**a**) ASF-SF cases at $x \approx -0.147D$, (**b**) RSF-SF cases at $x \approx -0.147D$, (**c**) ASF-SF cases at $x = 0$ and (**d**) RSF-SF cases $x = 0$.

It is worth noting that amplification factors of $\Gamma = 0.28$ and $\Gamma = 0.51$ are considered as 1.35 and 1.65, respectively, by computing U_H/U_∞. The amplification factor is defined as the ratio of wind speed in the case where there are buildings to wind speed in the case where these buildings are removed.

The separated shear flow from a building with an aspect ratio of 1:1:2 reaches an amplification factor of 1.2 [35]. In addition, the separated shear flow from a building with an aspect ratio of 1:1:6 reaches an amplification factor of 1.7 [36]. At $x = 0$, the values of Γ do no change, as shown in Figure 6c,d. However, due to the momentum diffusion, the horizontal gradients of the time-mean streamwise velocity become weaker. One of the reasons for the relatively large discrepancies between the profiles obtained by the experiment and the CFDs can be that the turbulence intensities of the CFDs are significantly small compared to those of the experiments, as shown in Figure 7. Even though very high values of turbulence intensity were set at the inlet boundary, the turbulence intensities dissipated rapidly and became very small at $x = 0$ as compared to those of the experiments. As the setting of high turbulence intensity at the inlet boundary leads to computational instability, we set no perturbation condition at the inlet boundary. Table 4 shows the values of U_0 computed by Equation (8) and the Reynolds number which is defined as:

$$Re = \frac{U_0 D}{\nu} \tag{18}$$

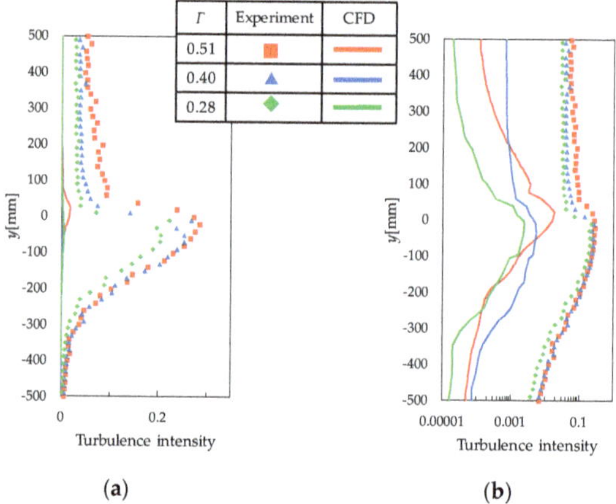

Figure 7. Horizontal distribution of the turbulence intensity of ASF-SF cases at the mid-height of the wind tunnel outlet at $x = 0$, which corresponds to the position of the rotational axis of the O-VAWT, when the O-VAWT was absent. (**a**) Linear scale on the horizontal axis and (**b**) logarithmic scale in the horizontal axis.

Table 4. Experimental and computational fluid dynamics (CFD) results for the values of U_0 and Re for the uniform flow, ASF-SF and RSF-SF cases.

Flow Type	Γ	U_0 [m/s]		Re	
		Experiment	**CFD**	**Experiment**	**CFD**
Uniform	0	8	8	2.79×10^5	2.79×10^5
ASF-SF	0.51	10.08	9.78	3.52×10^5	3.41×10^5
ASF-SF	0.40	9.38	9.22	3.28×10^5	3.22×10^5
ASF-SF	0.28	8.73	8.72	3.05×10^5	3.05×10^5
RSF-SF	0.51	7.76	7.48	2.71×10^5	2.61×10^5
RSF-SF	0.40	7.72	7.36	2.70×10^5	2.57×10^5
RSF-SF	0.28	7.52	7.33	2.62×10^5	2.56×10^5

At the outlet boundary, the pressure outlet condition with $p = 0$ was imposed. On the surface of the O-VAWT and the splitter plate, the no-slip boundary conditions were set. The sliding mesh technique was used to couple the rotational domains and the stationary domain. The direction of the rotor and the blade rotations are counterclockwise and clockwise, respectively, when viewed from the top (Figure 5a). By changing the rotational speed of the rotor ω, the tip speed ratio λ was set at 0.2, 0.4, 0.5, 0.6, or 0.8. The time step sizes were set as $dt = 0.5°/\omega$.

3.3. Torque and Power Coefficients

As mentioned in sub-Section 2.4, this study considers only the aerodynamic torque generated by the blades as the rotor torque of the O-VAWT. Since each of the blade axes was connected with the main shaft by a chain via sprockets, the aerodynamic torque on each of the blades about each of the blade axes was transmitted through the chain and contributed to the torque about the main shaft. Therefore, the rotor torque generated by a blade at an azimuthal angle φ is expressed as:

$$T_B(\varphi) = T_{B\_rev}(\varphi) + T_{B\_rot}(\varphi), \tag{19}$$

where $T_{B\_rev}(\varphi)$ is the conventional blade torque that is calculated by multiplying the rotor radius and the component of the aerodynamic force on the blade at φ in the rotor-revolution direction; and $T_{B\_rot}(\varphi)$ is the torque generated by the component of the aerodynamic force on the blade at φ in the blade-rotation direction about the blade axis. Hereafter, we call $T_B(\varphi)$ the "blade torque," $T_{B\_rev}(\varphi)$ the "rotor-revolution torque" and $T_{B\_rot}(\varphi)$ the "blade-rotation torque." The blade torque coefficient (C_{TB}), the rotor-revolution component (C_{TB_rev}) and the blade-rotation component (C_{TB_rot}) are defined as:

$$C_{TB}(\varphi) = \frac{T_B(\varphi)}{0.5\rho A U_0^2 R}, \tag{20}$$

$$C_{TB\_rev}(\varphi) = \frac{T_{B\_rev}(\varphi)}{0.5\rho A U_0^2 R}, \tag{21}$$

and

$$C_{TB\_rot}(\varphi) = \frac{T_{B\_rot}(\varphi)}{0.5\rho A U_0^2 R}. \tag{22}$$

It should be noted that C_{TB}, C_{TB_rev} and C_{TB_rot} are coefficients of one blade.

The CFD simulations were conducted for eight revolutions of the rotor. Using the data of the last two rotor revolutions, the torque coefficient C_T was computed by the following formula:

$$C_T = \frac{n}{N_e - N_s} \sum_{N=N_s+1}^{N_e} \int_0^{2\pi} C_{TB}\, d\varphi. \tag{23}$$

Here, $n\ (= 3)$ is the number of the blades; $N_s\ (= 6)$ is the number of the rotor revolutions before starting the computation of C_T; $N_e\ (= 8)$ is the number of the rotor revolutions before finishing the computation of C_T. The power coefficient of C_P was computed by Equation (6).

4. Results and Discussion

In this section, the results of the wind tunnel experiments and the CFD simulations for the power performance of the O-VAWT, such as the dependency of the power and torque coefficients on the tip speed ratio, the variations of the torque coefficients with azimuthal angle, are presented for the uniform flow case and the shear flow cases. Subsequently, the causes of the features of the power performance of the O-VAWT are discussed based on the CFD results of the flow fields.

4.1. Performance in Uniform Flow

Figure 8a shows the power and torque coefficients (C_P and C_T) of the O-VAWT in the uniform flow. The CFD results are in good agreement with the experimental ones. The optimal tip speed ratio at which C_P becomes the maximum is less than unity; 0.4 in the experiments and 0.5 in the

CFD simulations. As mentioned in the introduction, this low optimal tip speed ratio is a favorable feature for the built environment from the viewpoint of aerodynamic noise. With increasing λ, C_T decreases monotonically from a small tip speed ratio ($\lambda = 0.2$). These tendencies are commonly found among drag-type wind turbines. The wind-tunnel experimental results by Shimizu et al. [24] for C_P and C_T of an O-VAWT with two elliptical cross-sectional blades show the same tendencies as our results. The value of the maximum C_P is 0.32 in our experiments and CFD simulations, while the value is 0.176 in Shimizu et al.'s experiments. The main factors for the better performance of our O-VAWT as compared to Shimizu et al.'s O-VAWT can be the number of blades and the cross-sectional shape of the blades. In our previous studies, the maximum Cp improved from 0.189 to 0.244 by changing the number of blades from two to three [33] and improved from 0.246 to 0.288 by changing the cross-sectional shape of blades from ellipse to rectangle [23].

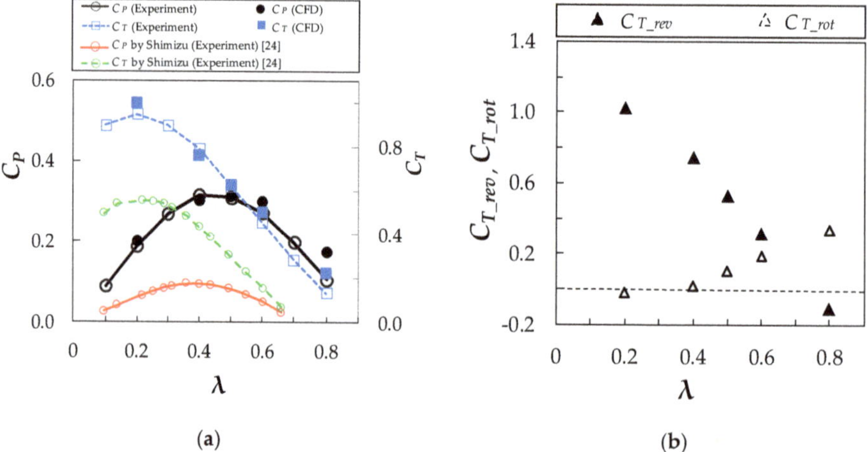

Figure 8. Performance of the O-VAWT in the uniform flow; (**a**) the variation of power coefficient (C_P) and torque coefficient (C_T) with tip speed ratios (λ) by the experiments and CFD simulations, (**b**) rotor-revolution (C_{T_rev}) and blade-rotation (C_{T_rot}) components of torque coefficients (C_T) computed by the CFD simulations. The wind-tunnel experimental results by Shimizu et al. [24] for C_P and C_T of an O-VAWT with two elliptical cross-sectional blades are added for reference.

Figure 8b shows C_{T_rev} and C_{T_rot} computed by the CFD results. The sum of C_{T_rev} and C_{T_rot} is C_T. As well as C_T, the value of C_{T_rev} decreases monotonically with an increase in λ. Conversely, the value of C_{T_rot} increases with an increase in λ. The values of C_{T_rev} are positive and larger than those of C_{T_rot} except for $\lambda = 0.8$. At $\lambda = 0.8$, C_{T_rev} is negative; however, C_{T_rot} is positive and its absolute value is larger than that of C_{T_rev}. As a result, C_T is positive at $\lambda = 0.8$.

Figure 9 shows the variations of blade torque coefficient (C_{TB}), its rotor-revolution component (C_{TB_rev}) and blade-rotation component (C_{TB_rot}) with respect to azimuthal angle (φ) at $\lambda = 0.4$ and 0.6. The value of C_{TB} is significantly large in the upwind region of the advancing side ($\varphi \approx 180°$ to $270°$) of the rotor, being the maximum at $\varphi \approx 210°$. In the range of φ where C_{TB} is significantly large, the contribution of C_{TB_rev} is dominant. Except for this range, C_{TB_rev} does not always positively contribute to C_{TB}. At φ where the value of C_{TB_rev} is negative, the value of C_{TB_rot} is generally positive and the rotation of the blade positively contributes to C_{TB}. Due to this positive contribution of C_{TB_rot} to C_{TB}, the value of C_{TB} is positive at almost all φ and the variation of C_{TB} of the O-VAWT with respect to φ is smaller as compared to that of a Savonius-type VAWT. The variation of C_{TB} of a Savonius-type VAWT with respect to φ in Figure 9 is a result of CFD simulation by Tian et al. [37]. The maximum C_P and the optimal λ of the Savonius-type VAWT were 0.258 and 1.0, respectively.

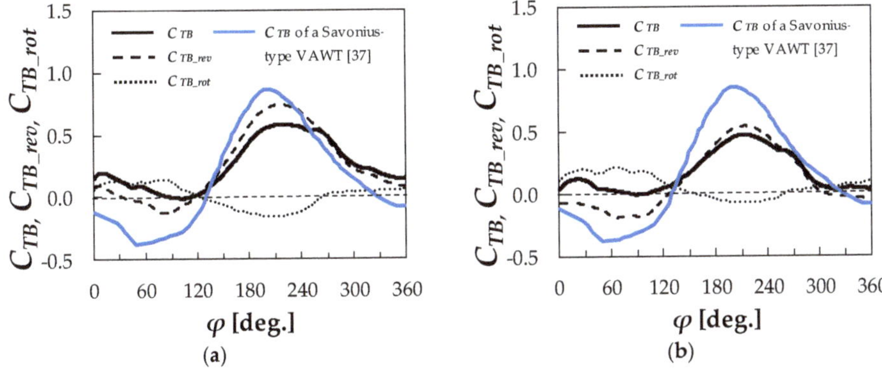

Figure 9. The variation of blade torque coefficients (C_{TB}), its rotor-revolution component (C_{TB_rev}) and its blade-rotation component (C_{TB_rot}) with azimuthal angle (φ) at; (**a**) $\lambda = 0.4$ and (**b**) $\lambda = 0.6$. Note that C_{TB}, C_{TB_rev} and C_{TB_rot} are coefficients of one blade. The CFD simulation result by Tian et al. [37] for the variation of C_{TB} of a Savonius-type VAWT with φ is added for comparison.

4.2. Performance in Shear Flows

Figure 10 compares the experimental results for C_P of the O-VAWT in the cases of the advancing side faster shear flow (ASF-SF), the retreating side faster shear flow (RSF-SF) and the uniform flow. In the cases of the ASF-SF, C_P is higher than in the case of the uniform flow at all λ and increases with an increase in Γ. In the cases of the RSF-SF, similar to the cases of the ASF-SF, C_P increases with an increase in Γ. However, when $\Gamma = 0.28$, the values of C_P are lower than those of the uniform flow case. Concerning the optimal λ, there is a trend that it shifts to higher λ in both cases of shear flows as compared to the uniform flow case, except for the RSF-SF with $\Gamma = 0.28$.

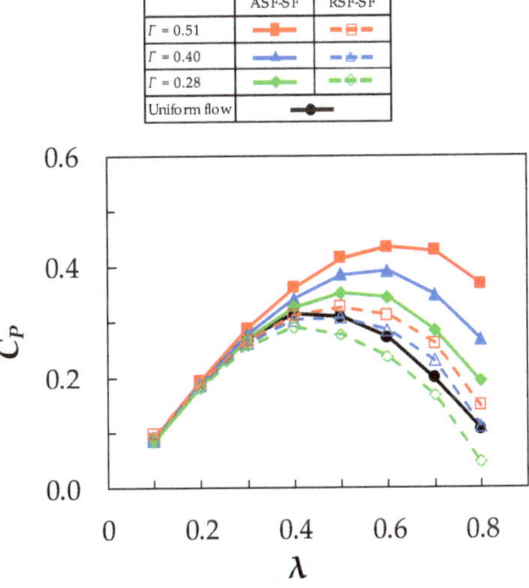

Figure 10. Power coefficient (C_P) of O-VAWT in shear flows by the experiments.

Both in the cases of the ASF-SF (Figure 11a) and the RSF-SF (Figure 11b), the CFD results for C_P–λ curves are in good agreement with the experimental ones. In the following discussion, we use the CFD results for the torque of the O-VAWT and the flow field to explain the effects of shear flows on the characteristics of the power performance.

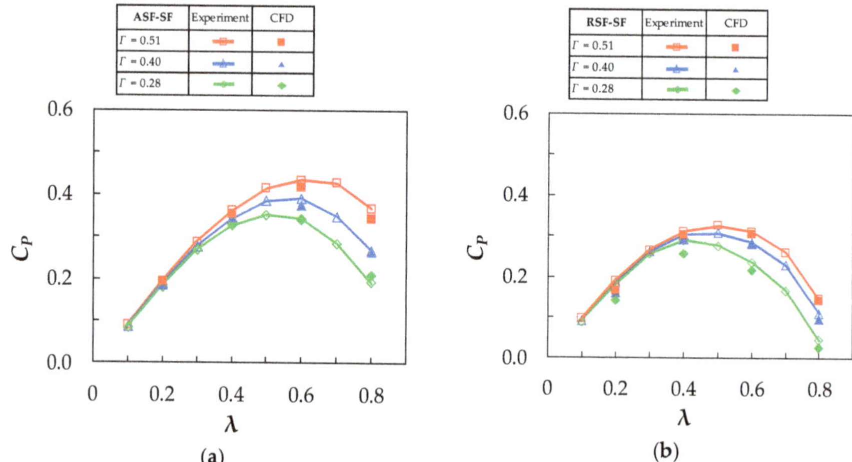

Figure 11. Power coefficient (C_P) of O-VAWT in shear flows by the experiments and the CFD simulations in the cases of; (**a**) the ASF-SF and (**b**) the RSF-SF.

Figure 12 shows the variations of blade torque coefficient (C_{TB}), its rotor-revolution component (C_{TB_rev}) and its blade-rotation component (C_{TB_rot}) with azimuthal angle (φ) in the cases of the ASF-SF. It is confirmed that these profiles are qualitatively the same between the cases of $\lambda = 0.4$ and $\lambda = 0.6$. Therefore, it is considered that the effects of the shear flow on the characteristics of the blade torque variations with φ do not significantly change around the optimal λ. As compared to the case of the uniform flow, C_{TB} is higher on most of the advancing side of the rotor ($\varphi \approx 210°$ to $330°$) and lower in the upwind region of the retreating side of the rotor, specifically at $\varphi \approx 120°$ to $150°$. In particular, with an increase in Γ, C_{TB} increases on most of the advancing side of the rotor ($\varphi \approx 210°$ to $330°$). The optimal φ at which C_{TB} is the maximum shifts to the downwind direction on the advancing side of the rotor ($\varphi \approx 240°$) as compared to the case of the uniform flow. The effects of the shear flow on the variations of C_{TB_rev} with φ is almost the same as C_{TB}. As compared to the case of the uniform flow, C_{TB_rev} is higher on most of the advancing side of the rotor ($\varphi \approx 210°$ to $330°$). In contrast, C_{TB_rot} is lower in most of the upwind region of the advancing side ($\varphi \approx 180°$ to $240°$) and slightly higher in the upwind region of the retreating side of the rotor, specifically at $120°$ to $150°$ as compared to the case of the uniform flow. With an increase in Γ, C_{TB_rot} decreases in the upwind region of the advancing side of the rotor ($\varphi \approx 180°$ to $240°$). Due to its negative values of C_{TB_rot}, the optimal φ at which C_{TB} is the maximum slightly shifts to the downwind direction as compared to C_{TB_rev}.

Figure 13 shows the variation of blade torque coefficient (C_{TB}), its rotor-revolution component (C_{TB_rev}) and its blade-rotation component (C_{TB_rot}) with azimuthal angle (φ) in the cases of the RSF-SF. Since these profiles are qualitatively the same between the cases of $\lambda = 0.4$ and $\lambda = 0.6$, it is considered that the effects of the shear flow on the characteristics of the blade torque variations with φ do not significantly change around the optimal λ. As compared to the case of the uniform flow, C_{TB} is significantly higher in most of the upwind region of the retreating side ($\varphi \approx 120°$ to $180°$) and lower on most of the advancing side of the rotor ($\varphi \approx 210°$ to $330°$). In particular, with an increase in Γ, C_{TB} increases in the upwind region of the retreating side of the rotor. The optimal φ at which C_{TB} is the maximum shifts to the upwind region of the retreating side of the rotor ($\varphi \approx 150°$) as compared to the case of the uniform flow. The effects of the shear flow on the variation of C_{TB_rev} with φ is almost the same as C_{TB}. As compared to the case of the uniform flow, C_{TB_rev} is higher in most of the upwind region of the retreating side ($\varphi \approx 100°$ to $160°$) and lower on most of the advancing side of the rotor ($\varphi \approx 210°$ to $330°$). In contrast, C_{TB_rot} is slightly lower in most of the upwind region of the retreating side ($\varphi \approx 90°$ to $150°$) and slightly higher in most of the upwind region of the advancing side of the rotor ($\varphi \approx 180°$ to $240°$) as compared to the case of the uniform flow. Furthermore, with an increase in Γ, C_{TB_rot} decreases in most of the upwind region of the retreating side ($\varphi \approx 100°$ to $160°$) and increases in most of the upwind region of the advancing side of the rotor ($\varphi \approx 180°$ to $240°$).

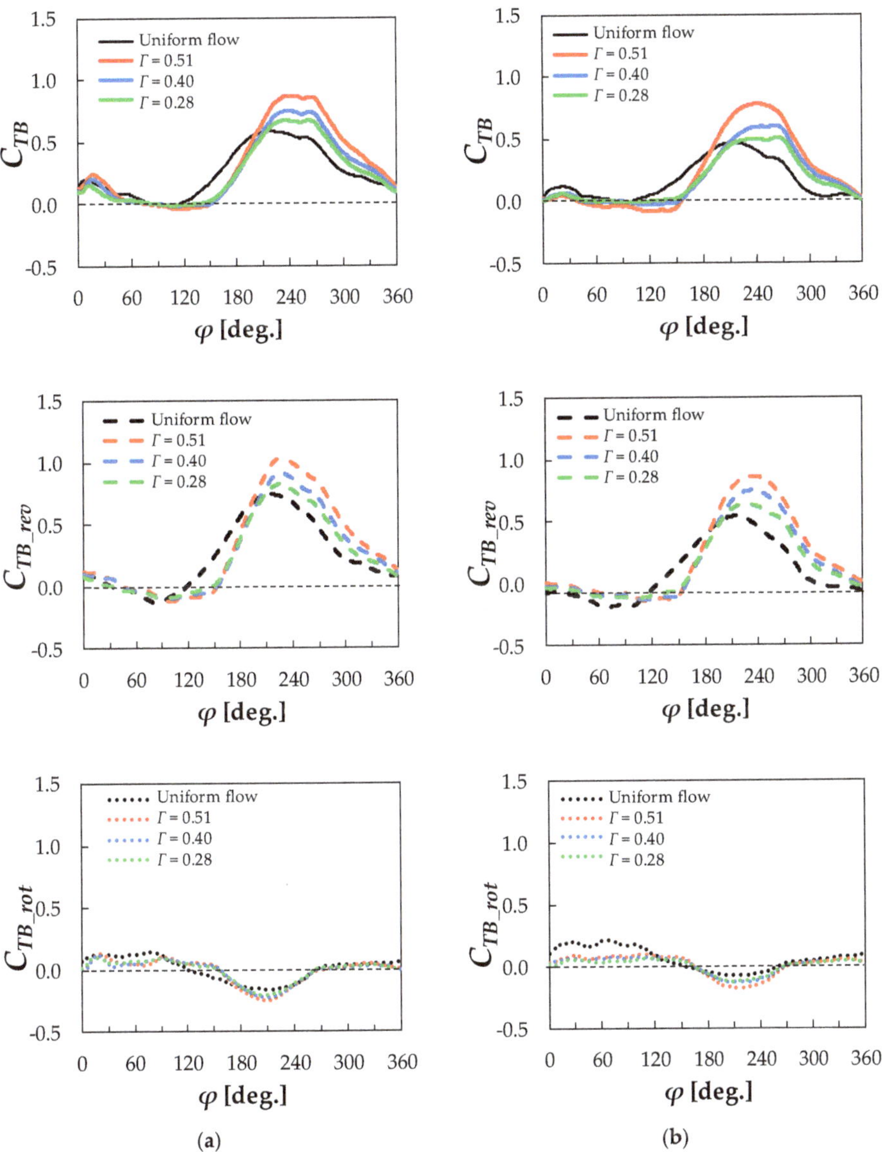

Figure 12. The variation of blade torque coefficient (C_{TB}), its rotor-revolution component (C_{TB_rev}) and its blade-rotation component (C_{TB_rot}) in the cases of the ASF-SF computed by the CFD simulations; (**a**) for $\lambda = 0.4$ and (**b**) for $\lambda = 0.6$. Note that C_{TB}, C_{TB_rev} and C_{TB_rot} are coefficients of one blade.

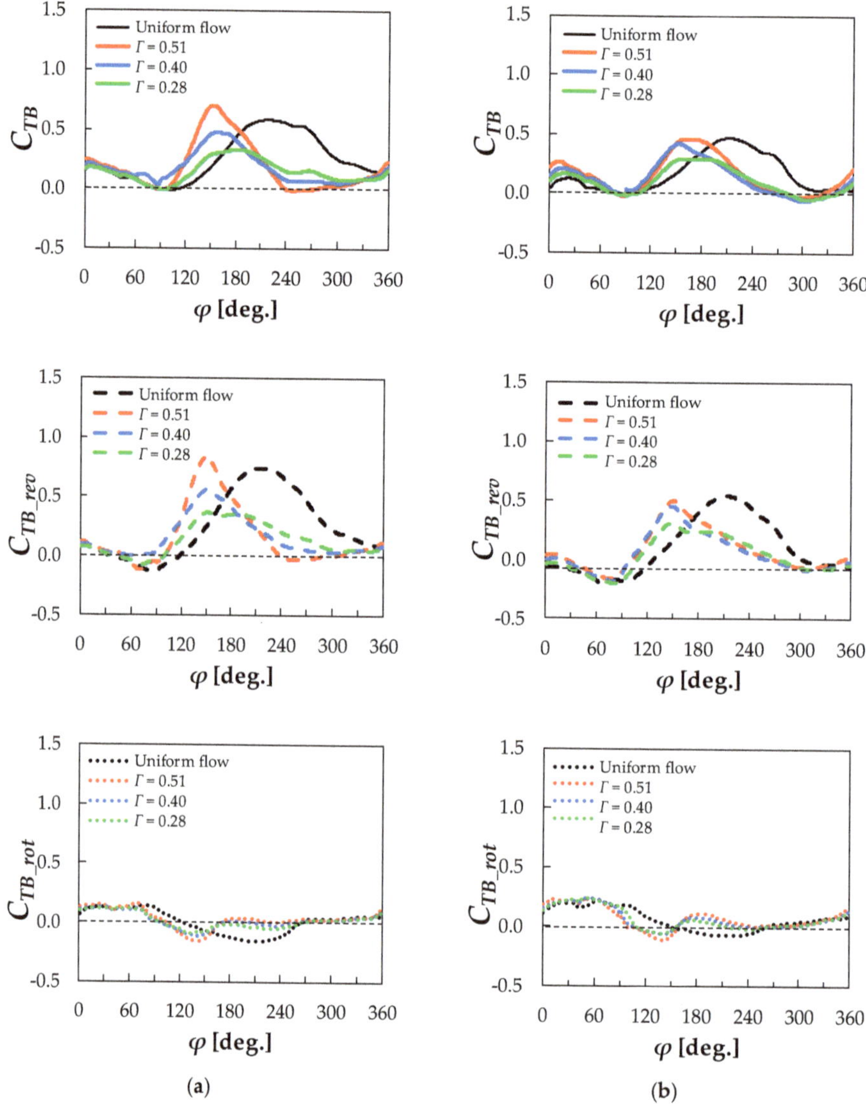

Figure 13. The variation of blade torque coefficient (C_{TB}), its rotor-revolution component (C_{TB_rev}) and its blade-rotation component (C_{TB_rot}) in the cases of the RSF-SF computed by the CFD simulations; (**a**) for $\lambda = 0.4$ and (**b**) for $\lambda = 0.6$. Noted that C_{TB}, C_{TB_rev} and C_{TB_rot} are coefficients of one blade.

4.3. Flow Characteristics

Figures 14a, 15a and 16a show the temporal sequence of the horizontal distributions of normalized horizontal velocity vectors, normalized pressure and normalized vorticity, respectively, at the mid-height of the O-VAWT at $\lambda = 0.4$ in the case of the uniform flow. At $\varphi = 180°$ to $270°$, the approaching flow to the blade has a large velocity component perpendicular to the blade, and the pressure on the upwind side of the blade is high. Due to this high pressure, C_{TB_rev} is significantly high in the range of $\varphi = 180°$ to $270°$ in Figure 9. In addition, at $\varphi = 210°$ to $270°$, due to the strong large vortex formed near the outer edge of the downwind side of the blade, the pressure is low near the vortex. This low pressure positively contributes to C_{TB_rev} (see Figure 9 at $\varphi = 210°$ to $270°$). By contrast, this low pressure negatively contributes to C_{TB_rot} (see Figure 9 at $\varphi = 210°$ to $270°$). At $\varphi = 300°$ and $330°$, the approaching flow to the blade has a small velocity component perpendicular to the blade and the pressure on the upwind side of the blade is not high. Therefore, C_{TB_rev} is lower as compared to the upwind region of the advancing side of the rotor (see Figure 9 at $\varphi = 180°$ to $270°$). At $\varphi = 0°$ to $120°$, the attack angle of the blade is positive (here, the counterclockwise direction is defined as positive) and the flow separates over the outer side of the blade. Therefore, on the upwind edge of the blade and on the inner side of the blade near its upwind edge, the pressure is relatively high. In contrast, on the outer side of the blade near its upwind edge, the pressure is relatively lower. This pressure distribution contributes negatively to C_{TB_rev} and positively to C_{TB_rot}.

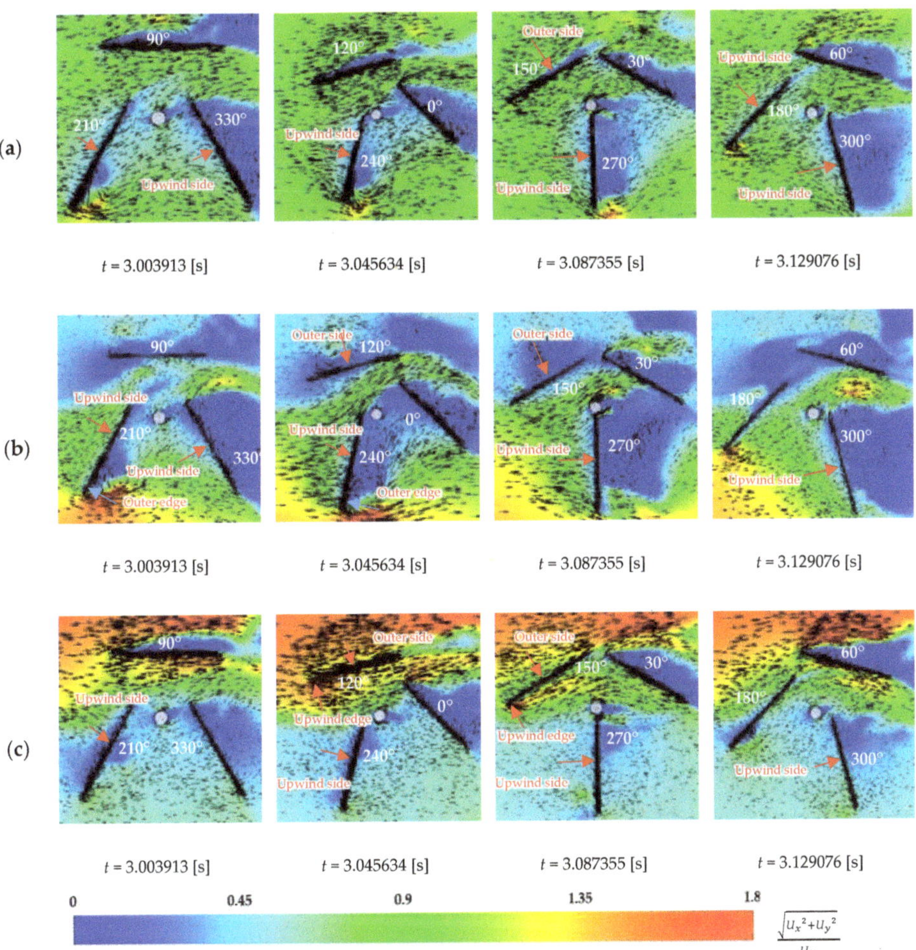

Figure 14. The temporal sequence of the horizontal distributions of normalized horizontal velocity vectors of the O-VAWT for $\lambda = 0.4$ in the case of; (**a**) the uniform flow, (**b**) the ASF-SF with $\Gamma = 0.51$ and (**c**) the RSF-SF with $\Gamma = 0.51$.

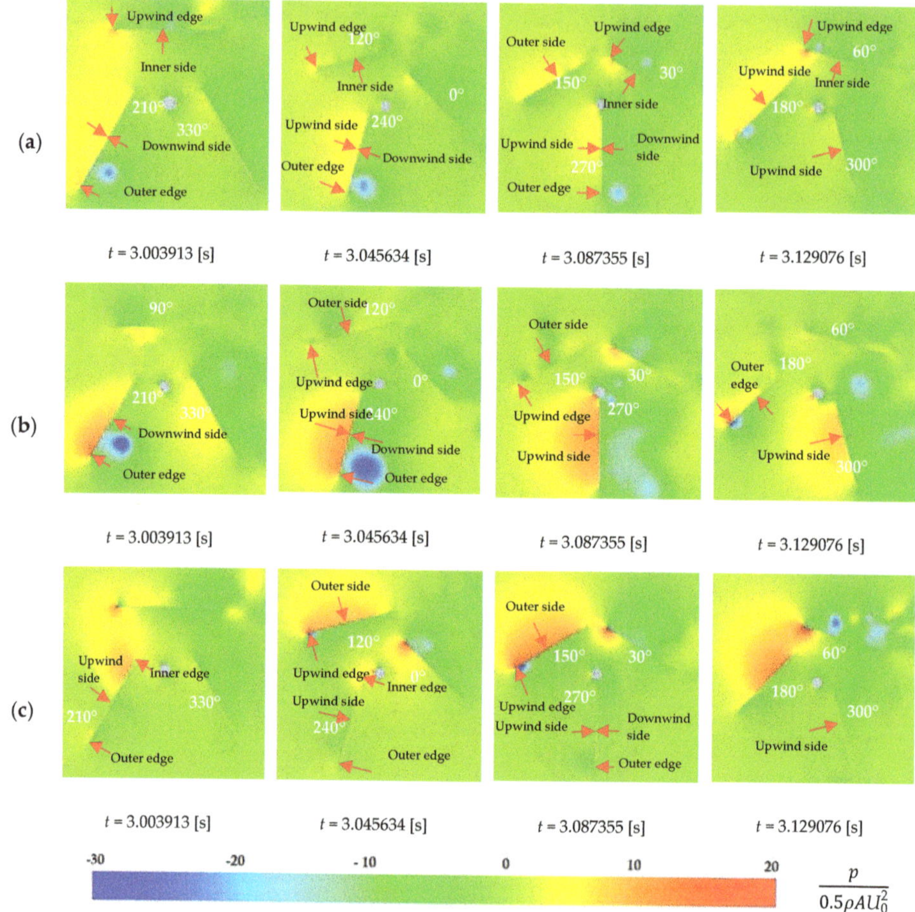

Figure 15. The temporal sequence of normalized pressure at the mid-height of the O-VAWT for $\lambda = 0.4$ in the case of; (**a**) the uniform flow, (**b**) the ASF-SF with $\Gamma = 0.51$ and (**c**) the RSF-SF with $\Gamma = 0.51$.

Figures 14b and 15b and Figure 16b show the temporal sequence of the horizontal distributions of normalized horizontal velocity vectors, normalized pressure and normalized vorticity, respectively, at the mid-height of the O-VAWT at $\lambda = 0.4$ in the case of the ASF-SF with $\Gamma = 0.51$. At $\varphi = 210°$ to $330°$, the approaching flow to the blade has a larger velocity component perpendicular to the blade, and the pressure on the upwind side of the blade is higher as compared to the uniform flow case. Due to this higher pressure, C_{TB_rev} is higher than the uniform flow case (see Figure 12 at $\varphi = 210°$ to $330°$). In addition, at $\varphi = 210°$ to $240°$, wind speed is significantly increased at the outer edge of the blade and a significantly stronger and larger vortex is formed near the outer edge on the downwind side of the blade. Near the vortex, the pressure is lower as compared to the uniform flow case. This low pressure contributes to higher C_{TB_rev} and lower C_{TB_rot} as compared to the uniform flow case (see Figure 12 at $\varphi = 210°$ to $240°$). At $\varphi = 120°$ and $150°$, the pressure on the outer side of the blade is lower due to the lower speed approaching flow to the blade. Furthermore, the pressure on the inner side of the blade near its upwind edge is higher due to the existence of the blade at $\varphi = 240°$ or $270°$, respectively. This pressure distribution contributes to lower C_{TB_rev} and higher C_{TB_rot} as compared to the uniform flow case (see Figure 12 at $\varphi = 120°$ to $150°$).

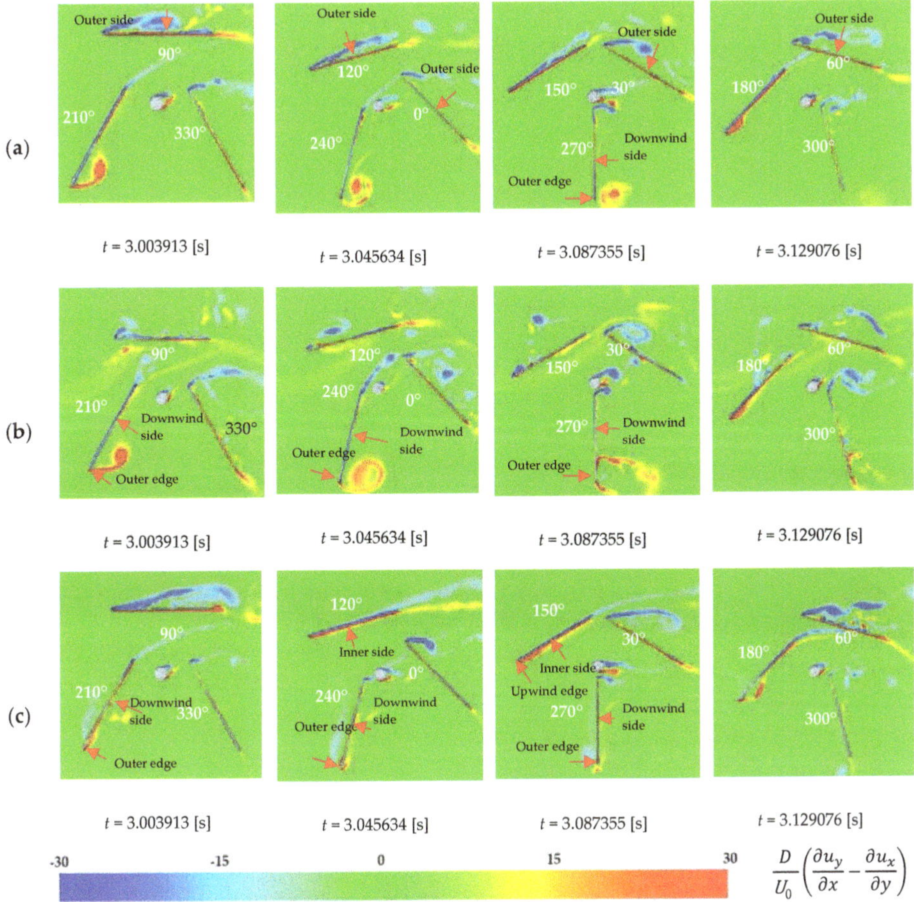

$$\frac{D}{U_0}\left(\frac{\partial u_y}{\partial x}-\frac{\partial u_x}{\partial y}\right)$$

Figure 16. The temporal sequence of normalized vorticity at the mid-height of the O-VAWT for $\lambda = 0.4$ in the case of; (**a**) the uniform flow, (**b**) the ASF-SF with $\Gamma = 0.51$ and (**c**) the RSF-SF with $\Gamma = 0.51$.

Figures 14c, 15c and 16c show the temporal sequence of normalized horizontal distributions of horizontal velocity vectors, pressure and vorticity, respectively, at the mid-height of the O-VAWT at $\lambda = 0.4$ in the case of the RSF-SF with $\Gamma = 0.51$. At $\varphi = 120°$ and $150°$, the approaching flow to the blade has a larger velocity component perpendicular to the blade and the pressure on the outer side of the blade is higher as compared to the uniform flow case. Due to this higher pressure, C_{TB_rev} is higher than the uniform flow case (see Figure 13 at $\varphi = 120°$ and $150°$). Furthermore, wind speed is increased at the upwind edge of the blade and a stronger vortex is formed near the upwind edge of the blade on its inner side. Near the vortex, the pressure is lower as compared to the uniform flow case. This low pressure contributes to higher C_{TB_rev} and lower C_{TB_rot} as compared to the uniform flow case (see Figure 13 at $\varphi = 120°$ and $150°$). At $\varphi = 210°$ to $300°$, except in the vicinity of the upwind side of the inner edge of the blade at $\varphi = 210°$, the approaching flow to the blade has a smaller velocity component perpendicular to the blade and the pressure on the upwind side of the blade is lower as compared to the uniform flow case. Due to this lower pressure, C_{TB_rev} is lower than the uniform flow case (see Figure 13 at $\varphi = 210°$ and $300°$). In addition, at $\varphi = 210°$ and $240°$, the vortex formed near the outer edge of the downwind side of the blade is weaker and the pressure drop becomes smaller as compared to the uniform flow case. This smaller pressure drop contributes to lower C_{TB_rev} and higher C_{TB_rot} as compared to the uniform flow (see Figure 13 at $\varphi = 210°$ and $240°$).

5. Conclusions

We investigated the effects of horizontal shear flow on the performance characteristics of an orthopter-type vertical axis wind turbine (O-VAWT) by conducting wind tunnel experiments and computational fluid dynamics (CFD) simulations. In addition to a uniform flow, two types of shear flow were used as the approaching flow to the O-VAWT. One type was an advancing side faster shear flow (ASF-SF), which had a higher velocity on the advancing side of the rotor. The other type was a retreating side faster shear flow (RSF-SF), which had a higher velocity on the retreating side of the rotor. For each type of shear flow, we set three different velocity ratios ($\Gamma = 0.28$, 0.40 and 0.51), which were the ratios of the difference between the highest velocity and the lowest velocity in a shear flow to the sum of the highest and lowest velocities. The main findings are summarized as follows:

1. In the ASF-SF cases, the power coefficients (C_P) were significantly higher than the uniform flow case at all tip speed ratios (λ) and increased with Γ. The experimental results for the maximum C_P of the ASF-SF case with $\Gamma = 0.51$ and the uniform flow case were 0.43 when $\lambda = 0.6$ and 0.32 when $\lambda = 0.4$, respectively. Around the optimal λ, the blade torque coefficient (C_{TB}) on the advancing side of the rotor was, in general, significantly higher than the uniform flow case and increased with Γ, predominantly contributing to the increase in C_P. The CFD results for the maximum discrepancies of C_{TB} on the advancing side of the rotor between the ASF-SF case with $\Gamma = 0.51$ and the uniform flow case were 0.38 when $\lambda = 0.4$ and 0.44 when $\lambda = 0.6$. The high values of C_{TB} of the ASF-SF cases on the advancing side of the rotor were mainly caused by the higher pressure on the upwind side of the blade due to the higher speed of the approaching flow and by the lower pressure near the outer edge of the downwind side of the blade due to the formation of a larger vortex.
2. In the RSF-SF cases, C_P increased with Γ. However, when $\Gamma = 0.28$, C_P was lower than the uniform flow case at all λ. When $\Gamma = 0.51$, C_P was higher than the uniform flow case except at low λ; however, it was lower than the ASF-SF case with $\Gamma = 0.28$. The experimental results for the maximum C_P of the RSF-SF case with $\Gamma = 0.28$, the RSF-SF case with $\Gamma = 0.51$ and the ASF-SF case with $\Gamma = 0.28$ were 0.29 when $\lambda = 0.4$, 0.33 when $\lambda = 0.5$ and 0.35 when $\lambda = 0.5$, respectively. Around the optimal λ, the blade torque coefficient (C_{TB}) on the retreating side of the rotor was, in general, higher than the uniform flow case and increased with Γ, predominantly contributing to the increase in C_P. The CFD results for the maximum discrepancies of C_{TB} on the retreating side of the rotor between the RSF-SF case with $\Gamma = 0.51$ and the uniform flow case were 0.60 when $\lambda = 0.4$ and 0.36 when $\lambda = 0.6$. The high values of C_{TB} of the RSF-SF cases on the retreating side of the rotor were mainly caused by the higher pressure on the outer side of the blade on the upwind side of the rotor, due to the higher speed of the approaching flow. By contrast, C_{TB} on the advancing side of the rotor was, in general, lower than the uniform flow case, due to the lower pressure on the upwind side of the blade.
3. C_{TB} consists of the rotor-revolution component (C_{TB_rev}) and the blade-rotation component (C_{TB_rot}). In all the shear flow cases, as well as the uniform flow case, the contributions of C_{TB_rev} to C_{TB} were dominant. The dependencies of C_{TB_rev} and C_{TB_rot} on Γ had the opposite tendencies.

These findings are useful for micro-siting of an O-VAWT in the area where shear flows occur. A location where ASF-SFs with high Γ values dominantly occur is ideal for installing the O-VAWT. At a location where not only ASF-SFs but also RSF-SFs occur at high frequencies, higher Γ values are preferable. However, the shear flows utilized in this study are limited in their profiles and the relative positions to the rotor. To properly conduct the micro-siting of an O-VAWT in the area where various kinds of shear flows occur, such as the vicinity of a building, it is essential to understand the performance characteristics of the O-VAWT in the various kinds of shear flows. Therefore, in future research, we plan to investigate the effects of the broadness of the shear layer and the relative position of the shear flow to the rotor on the O-VAWT's performance characteristics. In addition, we plan to investigate the effects of the turbulence intensity of the approaching flow on the CFD results for the O-VAWT's performance characteristics by setting obstacles, which emit eddies, upwind of the O-VAWT to avoid the rapid dissipation of high turbulence intensity.

Author Contributions: R.P.W. performed the numerical simulations and prepared this manuscript being supervised by T.K. (Takaaki Kono). All authors contributed to the analyses of the data. T.K. (Takaaki Kono) and T.K. (Takahiro Kiwata) supervised the entire work. All authors have read and agreed to the published version of the manuscript.

Funding: This research was funded by World Bank Loan No. 8245-ID and a grant from Takahashi Industrial and Economic Research Foundation.

Acknowledgments: This research was supported by the Program for Research and Innovation in Science and Technology (RISET-Pro) Kemenristekdikti. The authors are thankful to technician Kuratani and student Taiki Sugawara for their help with the experiment.

Conflicts of Interest: The authors declare no conflict of interest.

References

1. Bahaj, A.S.; Myers, L.; James, P.A.B. Urban energy generation: Influence of micro-wind turbine output on electricity consumption in buildings. *Energy Build.* **2007**, *39*, 154–165. [CrossRef]

2. Toja-Silva, F.; Colmenar-Santos, A.; Castro-Gil, M. Urban wind energy exploitation systems: Behaviour under multidirectional flow conditions—Opportunities and challenges. *Renew. Sustain. Energy Rev.* **2013**, *24*, 364–378. [CrossRef]

3. Ishugah, T.F.; Li, Y.; Wang, R.Z.; Kiplagat, J.K. Advances in wind energy resource exploitation in urban environment: A review. *Renew. Sustain. Energy Rev.* **2014**, *37*, 613–626. [CrossRef]

4. Tummala, A.; Velamati, R.K.; Sinha, D.K.; Indraja, V.; Krishna, V.H. A review on small scale wind turbines. *Renew. Sustain. Energy Rev.* **2016**, *56*, 1351–1371. [CrossRef]

5. Kumar, R.; Raahemifar, K.; Fung, A.S. A critical review of vertical axis wind turbines for urban applications. *Renew. Sustain. Energy Rev.* **2018**, *89*, 281–291. [CrossRef]

6. Stathopoulos, T.; Alrawashdeh, H.; Al-Quraan, A.; Blocken, B.; Dilimulati, A.; Paraschivoiu, M.; Pilay, P. Urban wind energy: Some views on potential and challenges. *J. Wind Eng. Ind. Aerodyn.* **2018**, *179*, 146–157. [CrossRef]

7. Toja-Silva, F.; Kono, T.; Peralta, C.; Lopez-Garcia, O.; Chen, J. A review of computational fluid dynamics (CFD) simulations of the wind flow around buildings for urban wind energy exploitation. *J. Wind Eng. Ind. Aerodyn.* **2018**, *180*, 66–87. [CrossRef]

8. Anup, K.C.; Whale, J.; Urmee, T. Urban wind conditions and small wind turbines in the built environment: A review. *Renew. Energy* **2019**, *131*, 268–283.

9. Carbó Molina, A.; De Troyer, T.; Massai, T.; Vergaerde, A.; Runacres, M.C.; Bartoli, G. Effect of turbulence on the performance of VAWTs: An experimental study in two different wind tunnels. *J. Wind Eng. Ind. Aerodyn.* **2019**, *193*, 103969. [CrossRef]

10. Wood, D. *Small Wind Turbines*; Green Energy and Technology; Springer: London, UK, 2011; ISBN 978-1-84996-174-5.

11. Manwell, J.F.; McGowan, J.G.; Rogers, A.L. *Wind Energy Explained: Theory, Design and Application*; John Wiley & Sons: Hoboken, NJ, USA, 2010; ISBN 9780470015001.

12. Allen, S.R.; Hammond, G.P.; McManus, M.C. Prospects for and barriers to domestic micro-generation: A United Kingdom perspective. *Appl. Energy* **2008**, *85*, 528–544. [CrossRef]

13. Islam, M.; Ting, D.S.K.; Fartaj, A. Aerodynamic models for Darrieus-type straight-bladed vertical axis wind turbines. *Renew. Sustain. Energy Rev.* **2008**, *12*, 1087–1109. [CrossRef]

14. Jin, X.; Zhao, G.; Gao, K.; Ju, W. Darrieus vertical axis wind turbine: Basic research methods. *Renew. Sustain. Energy Rev.* **2015**, *42*, 212–225. [CrossRef]

15. Ghasemian, M.; Ashrafi, Z.N.; Sedaghat, A. A review on computational fluid dynamic simulation techniques for Darrieus vertical axis wind turbines. *Energy Convers. Manag.* **2017**, *149*, 87–100. [CrossRef]

16. Kumar, P.M.; Sivalingam, K.; Narasimalu, S.; Lim, T.-C.; Ramakrishna, S.; Wei, H. A review on the evolution of darrieus vertical axis wind turbine: Small wind turbines. *J. Power Energy Eng.* **2019**, *7*, 27–44. [CrossRef]

17. Abraham, J.P.; Plourde, B.D.; Mowry, G.S.; Minkowycz, W.J.; Sparrow, E.M. Summary of Savonius wind turbine development and future applications for small-scale power generation. *J. Renew. Sustain. Energy* **2012**, *4*, 042703. [CrossRef]
18. Akwa, J.V.; Vielmo, H.A.; Petry, A.P. A review on the performance of Savonius wind turbines. *Renew. Sustain. Energy Rev.* **2012**, *16*, 3054–3064. [CrossRef]
19. Roy, S.; Saha, U.K. Review on the numerical investigations into the design and development of Savonius wind rotors. *Renew. Sustain. Energy Rev.* **2013**, *24*, 73–83. [CrossRef]
20. Kang, C.; Liu, H.; Yang, X. Review of fluid dynamics aspects of Savonius-rotor-based vertical-axis wind rotors. *Renew. Sustain. Energy Rev.* **2014**, *33*, 499–508. [CrossRef]
21. Alom, N.; Saha, U.K. Evolution and progress in the development of savonius wind turbine rotor blade profiles and shapes. *J. Sol. Energy Eng. Trans. ASME* **2019**, *141*, 1–15. [CrossRef]
22. Elkhoury, M.; Kiwata, T.; Nagao, K.; Kono, T.; ElHajj, F. Wind tunnel experiments and Delayed Detached Eddy Simulation of a three-bladed micro vertical axis wind turbine. *Renew. Energy* **2018**, *129*, 63–74. [CrossRef]
23. Kiwata, T. Vertical axis wind turbine with variable-pitch straight blades. In Proceedings of the International Conferemce on Jets, Wakes and Separated Flows, ICJWSF-2017, Cincinnati, OH, USA, 9–12 October 2017; pp. 1–6.
24. Shimizu, Y.; Maeda, T.; Kamada, Y.; Takada, M.; Katayama, T. Development of micro wind turbine (Orthoptere Wind Turbine). In Proceedings of the International Conference on Fluid Engineering JSME Centennial Grand Congress, Tokyo, Japan, 13–16 July 1997; pp. 1551–1556.
25. Bayeul-Laine, A.-C.; Simonet, S.; Dockter, A.; Bois, G. Numerical study of flow stream in a mini VAWT with relative rotating blades. In Proceedings of the 22nd International Symposium on Transport Phenomena Conference, Delft, The Netherlands, 8–11 November 2011; p. 13.
26. Cooper, P.; Kennedy, O.C.; Cooper, P.; Kennedy, O. Development and analysis of a novel vertical axis wind turbine development and analysis of a novel vertical axis wind turbine. In Proceedings of the Solar 2004: Life, The Universe and Renewables, Perth, Australia, 30 November–3 December 2004; pp. 1–9.
27. Kono, T.; Kogaki, T.; Kiwata, T. Numerical investigation of wind conditions for roof-mounted wind turbines: Effects of wind direction and horizontal aspect ratio of a high-rise cuboid building. *Energies* **2016**, *9*, 907. [CrossRef]
28. Balduzzi, F.; Bianchini, A.; Ferrari, L. Microeolic turbines in the built environment: Influence of the installation site on the potential energy yield. *Renew. Energy* **2012**, *45*, 163–174. [CrossRef]
29. Ferreira, C.J.S.; Van Bussel, G.J.W.; Van Kuik, G.A.M. Wind tunnel hotwire measurements, flow visualization and thrust measurement of a VAWT in skew. *J. Sol. Energy Eng. Trans. ASME* **2006**, *128*, 487–497. [CrossRef]
30. Burton, T.; Jenkins, N.; Sharpe, D.; Bossanyi, E. *Wind Energy Handbook*, 2nd ed.; John Wiley & Sons: Hoboken, NJ, USA, 2011; ISBN 9780470699751.
31. ANSYS, Inc. *ANSYS Fluent Theory Guide, Release 17.2*; ANSYS, Inc.: Canonsburg, PA, USA, 2016.
32. ANSYS, Inc. *ANSYS Fluent User Guide, Release 17.2*; ANSYS, Inc.: Canonsburg, PA, USA, 2016.
33. ElCheikh, A.; Elkhoury, M.; Kiwata, T.; Kono, T. Performance analysis of a small-scale orthopter-type vertical axis wind turbine. *J. Wind Eng. Ind. Aerodyn.* **2018**, *180*, 19–33. [CrossRef]
34. Spalart, P.R.; Jou, W.H.; Strelets, M.; Allmaras, S.R. Comments on the feasibility of LES for wings, and on a hybrid RANS/LES approach. In Proceedings of the First AFOSR International Conference on DNS/LES, Ruston, LA, USA, 4–8 August 1997; pp. 137–147.
35. Tominaga, Y.; Mochida, A.; Yoshie, R.; Kataoka, H.; Nozu, T.; Yoshikawa, M.; Shirasawa, T. AIJ guidelines for practical applications of CFD to pedestrian wind environment around buildings. *J. Wind Eng. Ind. Aerodyn.* **2008**, *96*, 1749–1761. [CrossRef]
36. Wu, H.; Stathopoulos, T. Wind-tunnel techniques for assessment of pedestrian-level winds. *J. Eng. Mech.* **1993**, *119*, 1920–1936. [CrossRef]
37. Tian, W.; Mao, Z.; Zhang, B.; Li, Y. Shape optimization of a Savonius wind rotor with different convex and concave sides. *Renew. Energy* **2018**, *117*, 287–299. [CrossRef]

MDPI

St. Alban-Anlage 66

4052 Basel

Switzerland

Tel. +41 61 683 77 34

Fax +41 61 302 89 18

www.mdpi.com

Applied Sciences Editorial Office

E-mail: applsci@mdpi.com

www.mdpi.com/journal/applsci